# Environmental Information Systems

Springer
*Berlin*
*Heidelberg*
*New York*
*Barcelona*
*Budapest*
*Hong Kong*
*London*
*Milan*
*Paris*
*Santa Clara*
*Singapore*
*Tokyo*

Oliver Günther

# Environmental Information Systems

With 82 Figures

 Springer

Professor Dr. Oliver Günther
Humboldt-Universität zu Berlin
Institut für Wirtschaftsinformatik
Spandauer Strasse 1
D-10178 Berlin, Germany

ISBN 3-540-60926-1 Springer-Verlag Berlin Heidelberg New York

Library of Congress Cataloging-in-Publication Data

```
Günther, Oliver, 1961-
    Environmental information systems / Oliver Günther.
       p.    cm.
    "Springer-Lehrbuch."
    Includes bibliographical references and index.
    ISBN 3-540-60926-1 (alk. paper)
    1. Information storage and retrieval systems--Environmental
sciences. 2. Environmental sciences--Data processing.    I. Title.
GE45.D37G86    1998
025.06'3637--dc21                                                98-5674
                                                                    CIP
```

© Springer-Verlag Berlin Heidelberg 1998
Printed in Germany

The use of general descriptive names, trademarks, etc. in this publication does not imply, even in the absence of a specific statement, that such names are exempt from the relevant protective laws and regulations and therefore free for general use.

Cover design: de'blik, Berlin
Typesetting: Camera-ready by author
SPIN: 10519726    45/3142 – 5 4 3 2 1 0 – Printed on acid-free paper

To Agnès, Juliette, and Amélie

# Preface

Environmental information systems are an exceptionally attractive application area for computer scientists because the societal impact of one's work is immediately apparent. While technical progress has been a key factor in the developments that led to the current environmental crisis, it is now essential to turn to technology once again to help us solve those problems. Inventing and developing computer technology for environmental applications is one important cornerstone of this endeavor.

There is a large number of ongoing projects to build or enhance environmental information systems in government and industry worldwide. Due to the complex nature of environmental data, these systems require state of the art computer technology, and users are often willing to experiment with new techniques. Technology transfer cycles are short, and interactions between users, vendors, and the research community are unusually intense.

Despite all these activities, there seem to be no textbooks on the subject of environmental information systems, with the exception of a few collections of articles. One reason may be that the field is developing fast and it is increasingly difficult to keep up with this pace during the time it takes to write a book. Moreover, the area is broad and inherently interdisciplinary. Anybody trying to cover this eclecticism is forced to take hard decisions on what to include and what to omit. This choice must be a personal one, and a responsible author is likely to feel frustrated about not being able to cover more of the subjects that relate to our field.

Nevertheless, I decided to write such a book, mainly because I think the field should be taught at a broader scale and that a textbook could help make this happen. Environmental information systems are an appropriate topic for a graduate class in a variety of subject areas, including environmental sciences, geography, geology, computer science, and business administration. I have taught such a class at Berkeley as a graduate course in computer science and at Humboldt as part of the graduate program in business administration and economics. In both cases there were a significant number of students from other departments, which made the classes a truly interdisciplinary experience.

A crucial question throughout was how to structure the book and thereby the field of environmental information systems. Up to now, the area is very application-oriented. There is a large number of system solutions under development and in practical use. On the other hand, there is no acknowledged conceptual framework into which these various contributions could be embedded. This book tries to provide such a framework by structuring the flow of environmental information into four phases: data capture, data storage, data analysis, and metadata management. This information flow corresponds to a complex aggregation process during which the incoming raw data is gradually transformed into concise documents that can be used for high-level decision support.

While including several real-life examples of environmental information systems in practice, the book still focuses on concepts rather than applications. Partly as a result of these considerations, several aspects of environmental information systems were kept rather short. In particular, the book contains only a brief treatment of environmental *management* information systems, i.e., systems to administer environmental information within the enterprise. This area has appeared only recently and it is still difficult to distinguish basic concepts from application- and country-specific features. The section on the World Wide Web concentrates on general principles how the Web can be used in the context of environmental information systems. It is not meant to provide an exhaustive list of relevant URLs, as the Web itself is a better place to generate and maintain such lists.

Two other subjects that were intentionally kept short are geographic information systems and simulation models. Both areas have a long tradition and there exists a large number of good textbooks and survey articles dedicated exclusively to them. I could not and did not want to compete with my colleagues in those fields and therefore decided to refer the reader to their works where appropriate.

Parts of this book are based on earlier publications that were co-authored with colleagues. Thanks in particular to Hans-Knud Arndt, Volker Gaede, Johannes Lamberts, Wolf-Fritz Riekert, and Agnès Voisard for their contributions and the years of collaboration that led up to them.

Eugene Wong at Berkeley and Terry Smith at Santa Barbara taught me many things about databases and geographic information systems before I became interested in environmental issues. In 1989, Franz Josef Radermacher offered me to head the environmental information systems group at FAW, an applied computer science research laboratory in Ulm (Germany). The four years in Ulm proved to be an extremely stimulating experience. Projects with Roland Mayer-Föll of the Baden-Württemberg Ministry of Environment and Traffic, and with other partners in government and industry were a constant source of ideas and, at the same time, an important measure of how practical our insights really were.

After moving to Berlin in 1993, projects with the German Federal Environmental Agency and the Lower Saxony Environmental Ministry played a similar role. Thanks to Helmut Lessing, Birgit Mohaupt-Jahr, Thomas

Schütz, Jürgen Seggelke, and Walter Swoboda for sharing their experiences with me and for telling me about the practical side of environmental data management.

Discussions with colleagues and students helped me to clarify many issues and to improve their presentation in this book. Without claiming completeness I would like to thank Dave Abel, Katja Andresen, Hemant Bhargava, Ralf Bill, Mathias Bock, Reinhold Burger, Andrew Frank, Mike Freeston, J. Christoph Freytag, Mike Goodchild, Ralf H. Güting, Hans-D. Haasis, Inge Henning, Günter Hess, Lorenz Hilty, Andreas Jaeschke, Thomas Kämpke, Andree Keitel, Ralf Kramer, Horst Kremers, Hans-Peter Kriegel, Ramayya Krishnan, Ralf-D. Kutsche, Peter Ladstätter, Hans-Joachim Lenz, Peter Lockemann, Peter Mieth, Manfred Müller, Rudolf Müller, Michael Mutz, Hartmut Noltemeier, Tamer Özsu, Alfred Ortmaier, Bernd Page, Claus Rautenstrauch, Jan Röttgers, Thomas Ruwwe, Hanan Samet, Hans-J. Schek, Carsten Schmidt, Knut Scheuer, Matthäus Schilcher, Carsten Schöning, Michel Scholl, Klaus-Peter Schulz, Heinz Schweppe, Eric Simon, Marcus Spies, Myra Spiliopoulou, Franz Steidler, Jürgen Symanzik, Bernhard Thalheim, Megan Thomas, Anthony Tomasic, Ulrich Verpoorten, Mark Wallace, Herbert Weber, Marvin White, Peter Widmayer, and Gerlinde Wiest.

I would also like to thank my hosts in Berkeley and Paris, where large parts of this book were written. Jerry Feldman, Joe Hellerstein, and Robert Wilensky helped to make my sabbatical in Berkeley a fruitful and pleasant experience. Ulrich Finger, Philippe Picouet, and Jean-Marc Saglio of the École Nationale Supérieure des Télécommunications were welcoming hosts and stimulating discussion partners whenever I was visiting them in Paris.

Research related to this book was funded by the German Research Society (DFG grant nos. SFB 373/A3 and GRK 316), the German Federal Environmental Agency, the Environmental Ministries of Baden-Württemberg and Lower Saxony, the Baden-Württemberg Environmental Protection Agency, the Baden-Württemberg Geological Survey, and the following companies: Digital Equipment GmbH, ESRI GmbH, IBM Deutschland GmbH, Hewlett-Packard GmbH, Siemens AG, Siemens Nixdorf Informationssysteme AG, strässle Informationssysteme GmbH. Additional travel grants were provided by the Alexander von Humboldt Foundation (TransCoop Program), the European Union (ESPRIT Research Initiative 8666 – CONTESSA), the German Academic Exchange Service (DAAD grant no. 312/pro-gg), and the International Computer Science Institute. I am appreciative of their support.

Cooperation with Springer-Verlag was excellent throughout the two years it took to write this book. J. Andrew Ross and Hans Wössner followed the project closely and provided many helpful suggestions. Frank Holzwarth's detailed knowledge of LaTeX was of great help to improve the book's layout.

Everybody who ever wrote a book knows that such a project tends to occupy one's mind, therefore taking time away from one's personal life. I would like to thank Agnès, Juliette, Amélie, our families, and our friends for their tolerance and understanding during this period.

Berlin, January 1998                                        Oliver Günther

# Table of Contents

# List of Figures

# List of Tables

# 1. Introduction

The preservation of the environment has become an important public policy goal throughout the world. Citizens are taking a greater interest in the current and future state of the environment, and many are adapting their way of life accordingly. Companies are required to report on the environmental impact of their products and activities. Governments are concerned more than ever about their environmental resources and are establishing policies to control their consumption. To devise and implement such policies, administrators require detailed information about the current state of the environment and ongoing developments.

Moreover, an increasing number of governments are starting to recognize the right of their citizens to access the environmental information available. According to a recent directive of the European Union, for example, almost all environmental data that is stored at public agencies has to be made available to any citizen on demand [Cou90]. As the last few years have shown, the tendency to exert this right is rising steadily. Citizens, special-interest groups, and government agencies alike are requesting up-to-date information regarding air and water quality, soil composition, and so on.

As a result of these political and economic developments, there is a major demand for environmental information and appropriate tools to manage it. Given the amount and complexity of environmental data, these new information needs can only be served by using state-of-the-art computer technology.

*Environmental information systems* are concerned with the management of data about the soil, the water, the air, and the species in the world around us. The collection and administration of such data is an essential component of any efficient environmental protection strategy. Vast amounts of data need to be available to decision makers, mostly (but not always) in some kind of condensed format. The requirements regarding the currency and accuracy of this information are high. Details vary between applications [SBM94, HPRR95]. While earth scientists, for example, often emphasize the need to support system modeling [Edd93], scientists from other disciplines may put their emphasis on powerful database systems [Gos94] or systems integration [SMN94].

This book describes the design and implementation of information systems to support decision-making in environmental management and protection. While the required information technology is rarely domain-specific, it is important to select and combine the right tools among those that are available in principle. This requires a thorough knowledge of related developments in computer science and a good understanding of the environmental management tasks at hand.

The first step in any kind of data processing, computer-based or not, concerns the mapping of real-world objects to entities that are somewhat more abstract and can be handled by computers or directly by the decision maker. In this book we refer to the real-world objects of interest as *environmental objects*. Just about any real-world object can be regarded as an environmental object. Note in particular that the term is not restricted to natural entities (such as animals or lakes), but also includes human-made structures (such as houses or factories).

Each environmental object is described by, or mapped to, a collection of *environmental data objects*. These objects are abstract entities that can be handled by computers or by decision makers directly. A typical environmental data object would be a series of measurements that captures the concentration of a certain substance in a river (the corresponding environmental object).

While it is important to distinguish a data object from its physical and visual representations, the boundaries are sometimes fuzzy. An environmental data object can be available in analog or digital form. In the past, environmental information systems have been analog. Early attemps to understand and to manage the environment are as old as civilization itself. They are documented in ancient maps, collections of measurements and observations, hunting schedules, and so on – the environmental data objects of that period. The computer has opened up a whole new range of instruments to manage environmental data. As a result, environmental data objects are increasingly becoming available in digital form, with the trend toward digitized maps just being the most visible sign of this transition. As a consequence of these developments, we see that certain entities that would traditionally have been regarded as autonomous data objects are only visualizations of other data objects. A specialized map, for example, is often just a view or presentation tool for the underlying data.

In this book we use the term *environmental data object* in a wide sense, oriented toward the users' perceptions. Whatever they perceive as a separate information entity is called an environmental data object. Nesting or overlaps between such objects are common.

Computerized environmental information systems are able to collect and process much greater amounts of data than anybody would have thought only a few years ago. Automatic data capture and measurement results in the processing of terabytes of data per day. Even in processed form, this kind of data is impossible to browse manually in order to find the information that is relevant for a given task. Modern information retrieval tools allow the automatic or semi-automatic filtering of the available data in order to find those data sets one is looking for. An important prerequisite for these tools is the availability of *metadata*, i.e., data about data.

In this book, we call the metadata that refers to a particular environmental data object its *environmental metadata*. Each environmental data object is typically associated with one or more *metadata objects* that specify its format and contents. The documentation of the measuring series described above would be a typical example. It may include data about the spatial and temporal scale of the measurements, the main objectives of the project, the responsible agency, and so on. As soon as an environmental data object is created or updated, the change should be propagated to the metadata level. From there it may be forwarded to any application for which this modification may be relevant.

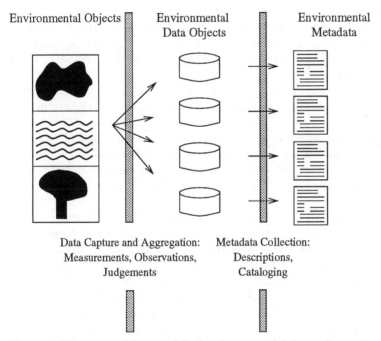

**Fig. 1.1.** Three-way object model of environmental information systems

These considerations lead to a three-way object model (Fig. 1.1). An environmental object is described by one or more environmental data objects. Each environmental data object may in turn be described by some environmental metadata. The data flow that is typically associated with such an approach closely resembles the data flow in classical business applications. It can be structured into four phases: data capture, data storage, data analysis, and metadata management.

1. The first phase, *data capture*, concerns the collection and processing of environmental raw data, such as measurement time series or aerial photographs. In this step the great variety of environmental objects is mapped onto a collection of environmental data objects, which usually have a structure that is much simpler and more clearly defined than the original raw data. There are a variety of ways to perform such a mapping, including measurement and observation, but also value-based judgement. Data capture usually involves some considerable aggregation, where the raw data is condensed and enriched in order to extract entities that are semantically meaningful. In the case of image data, for example, this includes the recognition of geometric primitives (such as lines and vertices) in an array of pixels, the comparison of the resulting geometric objects with available maps, and the identification of geographic objects (such as cities or rivers) in the picture. The information can then be represented in a much more compact format (in this case, a vector-based data format, as opposed to the original raster data). Measurement time series also need to be aggregated and possibly evaluated by means of some standard statistical procedures. The aggregated data is then stored in a file or a database.

2. For *data storage*, one has to choose a suitable database design and appropriate physical storage structures that will optimize overall system performance. Because of the complexity and heterogeneity of environmental data, this often necessitates substantial extensions to classical database technology. Non-standard data types and operators need to be accommodated and supported efficiently. In particular the management of spatial data requires the integration of highly specialized data structures and algorithms.

3. In the *data analysis* phase, the available information is prepared for decision support purposes. This may require simultaneous access to data that is geographically distributed, stored on heterogenous hardware, and organized along a wide variety of data models. Data analysis is typically based on complex statistical methods, scenarios, simulation and visualization tools, and institutional knowledge (such as environmental legislation or user objectives). Only the synthesis of the input data and these kinds of model allows us to judge the state of the environment and the potential of certain actions, both planned and already implemented. With regard to aggregation, data analysis can be seen as a direct continuation of data capture. The main difference is that the aggregation is now more target-oriented, i.e., more specifically geared towards particular tasks and decision makers.

4. *Metadata* is collected and aggregated throughout the three phases described above. It is stored in appropriate data structures and serves mostly in the data analysis phase to support search and browsing operations.

The overall objective of this complex aggregation process is to provide decision support at various levels of responsibility. Figure 1.2 uses the symbol of the pyramid to visualize this idea. Data aggregation corresponds to a bottom-up traversal of the central pyramid. Data can be used throughout that traversal for decision support purposes. While the data in the lower part of the pyramid tends to be used for local, tactical tasks, the upper part corresponds to strategic decision support for middle and upper management.

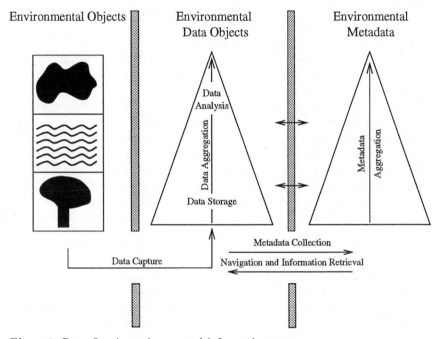

**Fig. 1.2.** Data flow in environmental information systems

Given the fact that in most Western industrial nations one can recognize major bottlenecks especially in data aggregation and analysis, it is critical that information systems technology is put to much wider use in environmental management. While there are many parallels between the kind of sequential data processing sketched above and the data flow in traditional business applications and geographic information systems (GIS), environmental applications often combine several properties that are problematic from a data management point of view:

– The *amount of data* to be processed is unusually large. The amount of satellite imagery recorded per day, for example, is already in the terabyte range [CC90]. That is about two orders of magnitude beyond the typical size of large high-transaction databases in banking or airline reservation applications. The processing of such large amounts of data demands hardware and software tools that reflect the state of the art in computer technology. Classical storage technology and data management techniques are no longer sufficient.

- Data management is usually highly *distributed*. Environmental data is captured, processed and stored by a broad range of government agencies and other institutions.
- The data management is extremely *heterogenous*, in terms of both hardware and software platforms. Data is organized according to a wide variety of data models, depending on the primary objectives of the particular agency in charge.
- Environmental data objects frequently have a *complex internal structure*, i.e., they consist of subobjects. These components may be complex objects in turn and may be associated with heterogenous types and media (including sound or images).
- Environmental data objects are often *spatio-temporal*, i.e., they have a location and a spatial extension, and they change over time.
- Environmental data is frequently *uncertain*. Techniques from statistics and from artificial intelligence have to be employed to manage this uncertainty.
- Because environmental issues cut across traditional subject areas, the processing of user queries may require *complex logical connections and joins*. Data is often used for purposes that are very different from the context the original data providers had intended.

After a detailed discussion of these problem areas, this book will present a variety of techniques that have been developed to handle these difficulties. While the emphasis of the book will be on questions that are of direct interest to the database and information systems communities, related issues from the areas of systems theory, artificial intelligence, and image processing will also be discussed. The structure of the book closely follows the data flow pictured in Fig. 1.2. One chapter each is devoted to data capture, data storage, data analysis, and metadata. Each chapter includes real-life examples taken from environmental information systems that are currently in use by government agencies and the general public.

Several aspects of environmental information systems were intentionally kept short. In particular, the book contains only a brief treatment of environmental *management* information systems, i.e., systems to administer environmental information within the enterprise. This area is of very recent origin and still very much in flux. In addition, many related issues are of an institutional character and therefore greatly depend on the given legislative and organizational framework. As a result, publications in this area are often specific to a particular country or branch of industry, and it is difficult to identify more general principles and paradigms. Nevertheless, there have been several attempts to survey the state of the art, and we will refer the reader to those articles for more information.

Our treatment of the World Wide Web focuses on general principles how the Web can be used in the context of environmental information systems. It is not meant to provide an exhaustive list of relevant URLs, as the Web itself is a better place to generate and maintain such lists. However, we will provide pointers to indices, search engines, and metainformation systems that can be used to search for environmentally relevant sites (cf. Sect. 4.4.2 and Chap. 5). In addition, we provide an appendix with all relevant URLs cited in this book.

Two other subjects that were kept short are geographic information systems and simulation models. Both areas have a long tradition and there are numerous good textbooks and survey articles dedicated exclusively to them. References to those works are included where appropriate. In this book, we restrict ourselves to those aspects that are most relevant in environmental applications. The use of simulation models in the environmental sciences is explained in Sect. 4.2, and GIS-related questions are discussed in Sect. 3.1, 3.4, and 4.3.

The book is appropriate as a textbook for a one-quarter or one-semester course with 30 to 50 hours of lecture time. If one operates at the lower end of this range, one may want to cut one of the two case studies in Chap. 2 (i.e., Sect. 2.4 or 2.5). One could also shorten the treatment of multidimensional access methods (Sect. 3.3) and object-oriented techniques (Sect. 3.4), and drop some of the case studies in Chap. 5.

# 2. Data Capture

Due to the political developments of the 1970s and 1980s, both the amount and the quality of the data that is being collected has increased considerably over the past 10 to 15 years. Environmental management and protection have become household terms. In the Western industrial nations, most governments are encouraged by their constituencies to pay attention to environmental issues and to support efforts to increase the efficiency of environmental protection. Green parties have been voted into numerous parliaments, and although they try to broaden their political agenda, environmental protection still forms their focus of activities and main area of expertise.

Two other developments have helped to turn this general political trend into concrete activities, in particular improvements in the collection, management, and utilization of environmental information. First, there is the ongoing industrial conversion resulting from the end of the cold war. As military budgets were cut, the demand for arms and other military technology dropped considerably. When looking for alternatives to make up for this major loss of business, many companies discovered the environmental sector as a potential market. Especially sensor technology could easily be converted to match the requirements of environmental applications. While one may feel cynical about this "peace dividend," there can be no doubt that it meant a considerable windfall profit for the environmental sector.

A second trend that proved to be crucial for environmental information management concerns the continued progress in computer technology during the past decade. Environmental data sets are large and often complex. Their administration requires powerful processors and efficient storage technology. Given the financial means available for environmental activites, efficient environmental information management would not have been possible on a large scale as recently as 1980. Since then the price/performance ratio of processors has dropped by several orders of magnitude, as have the prices for main memory and disk space. Governments in many Western industrial countries have reacted to these trends. Sensor networks have been installed and upgraded to monitor the quality of the water, the air, and the ground countrywide. Satellites are used increasingly to obtain environmental data, not only about remote areas. As a result, the availability of raw data about the environment is no longer a bottleneck, at least not for the Western industrial countries,

Australia, New Zealand, Japan, and the Tiger states of South East Asia.[1] The question is rather how to process these large amounts of data in order to obtain efficient decision support.

Because of the political and administrative developments described above, the amounts of data that are collected about the environment have grown extremely large. The environmental data collected per day worldwide already exceeds the 10 terabyte mark, and we can expect this to grow by one or two orders of magnitude by the end of the century. Most of the collected data sets are unstructured raw data, in particular raster images. They require a considerable amount of processing before they can be used for decision support purposes. Given the size of the data sets, it is not realistic to rely solely on human expertise in order to aggregate and evaluate this data. In many areas there are simply not enough human experts available, and even if they were, it would not make economic sense to employ them for the often mundane task of raw data processing. Usually the government agency (or the company) in charge is neither willing nor able to pay for this service. As a result, there are already many cases where raw data sets are written to disk without anybody ever having had a look at them. Once again, we see that raw data is not the bottleneck – *evaluation* is!

## 2.1 Object Taxonomies

The term data capture denotes the process of deriving environmental data objects from environmental objects. As noted in the introduction, just about any real-world object can be regarded as an environmental object. Environmental objects can be grouped into a number of classes. The *atmosphere* includes all objects above the surface of the Earth, such as the air or most kinds of radiation. The *hydrosphere* contains water-related objects, such as lakes or rivers. For frozen water such as snow and ice, one sometimes uses the term *cryosphere*. The term *lithosphere* relates to soil and rocks. The *biosphere* is the collection of all living matter, i.e., animals and plants. The term *technosphere* is used to denote human-made objects, such as houses or factories.

Figure 2.1 illustrates this object taxonomy and shows typical attributes for some of the classes. Note that, while in common use, the taxonomy is far from being unambiguous. An artificial lake, for example, is part of both hydrosphere and technosphere. A given land parcel may be described both as a part of the soil (lithospere) and as a piece of cultivated land.

---

[1] This is of course not true for most other countries. As we know now, the countries of the former Warsaw Pact did not put a major amount of resources into environmental management. Measuring networks are underdeveloped, if they exist at all. The situation is even worse for Africa, China, and large parts of South America.

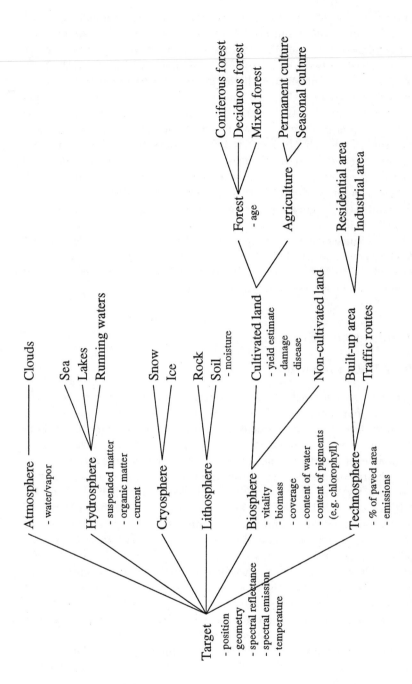

**Fig. 2.1.** Taxonomy of environmental objects [GHM+93, with kind permission from Kluwer Academic Publishers]

A simpler version of this taxonomy structures the environment into just three *media*: *ground, water*, and *air*. This taxonomy, which is commonly used by government agencies, leads to a relatively straightforward mapping to the traditional scientific disciplines. Geology, mineralogy, and parts of the agricultural sciences are concerned with the ground. Hydrology, oceanography, and limnology are about water, and the atmospheric sciences, including meteorology, are concerned with the air. Many public administrations are still organized along this taxonomy. As we begin to understand the environment as an integrated and multiply connected complex system, however, this taxonomy is found increasingly insufficient. Many natural processes and effects cannot be assigned to one of those three media alone. Understanding the environment is an inherently interdisciplinary task that transcends those traditional boundaries. Academic programs in the environmental sciences often reflect this fact, drawing from the course offerings of a wide range of departments. Administrative structures are starting to change as well, albeit more slowly. While one still finds the classical departments, especially when it comes to data capture, interdisciplinary task forces and project-oriented matrix organizations are becoming increasingly common.

## 2.2 Mapping the Environment

### 2.2.1 Raw Data Processing

Whatever taxonomy is used, the main question is which environmental objects should be monitored, and what data should be collected on them. There are many ways to obtain environmental data objects from environmental objects. One may use sensors to monitor, for example, the temperature of the air or the emissions of a factory. The results are archived as time series of measurements. More complex analytical techniques, such as chromatography or mass spectrometry, may be used to survey the chemical content of a lake. Air and satellite imagery is increasingly being used to monitor remote areas and to recognize long-term environmental developments. For that purpose, the raw imagery is usually processed and represented as a thematic map in order to visualize, for example, land utilization or temperature distribution. For forest and wildlife management, controlled observations and manual counts of animals or plants are often the most reliable source of data. For objects of the technosphere, it is frequently useful to study some written documentation in order to extract and condense the required environmental data objects.

In all of these cases, the incoming raw data first has to be subjected to some domain- and device-specific processing. Depending on the data source, this may include some optical rectification, noise suppression, filtering, or contrast enhancement.

Even at this early stage of data capture, it is not realistic to assume that this process can always be handled in an objective manner. Subjective influences, such as opinions and judgements of the people involved, often find their way into the data capture process. This may happen unknowingly or in full conscience (for instance, in order to judge the hazard potential of a factory).

### 2.2.2 Classification

If there is too much data for manual evaluation, the natural question is what parts of data capture could and should be automated. While the initial raw data processing, such as rectification and filtering, is usually performed directly by the sensor equipment, the following steps are somewhat more difficult to automate. A particularly important problem concerns the question how to classify a new observation with respect to a given taxonomy. The evaluation of satellite imagery provides a typical example. Given an unidentified pixel and a variety of possible land uses, the question is how to identify the most likely kind of land use for this pixel. Similar problems come up in other data capture situations as well.

The commonest technique for solving this problem is called *maximum likelihood*. This approach is based on the assumption that there exists a finite number of classes $\omega_i$ to which the new observation may belong. Each observation is represented as a point $X$ in some $n$-dimensional feature space. $X$ is sometimes referred to as the observation's *signature*. Furthermore, for each class $\omega_i$ there exists an $n$-dimensional probability distribution $p(X|\omega_i)$ that indicates the probability that a member of class $\omega_i$ assumes the observation value $X$. This probability distribution, however, is usually unknown. In most applications, one therefore assumes a normal distribution. As empirical studies have shown, slight violations of the normal distribution do not affect the classification accuracy in a major way [SD78].

Once a normal distribution has been adapted, the problem remains how to estimate the parameters of the distribution. Both the mean $\mu_i$ and the covariance matrix $\Theta_i$ have to be estimated for each class $\omega_i$. This is usually done using a suitable set of *training data*. The training data consists of observations whose class affiliation is known. In order to avoid singularities, one needs to take a representative sample of at least $n + 1$ observations. More observations improve the classification performance; in some applications a sample size of $10n$ is not uncommon [SD78]. Given a sample $X_{i1}$, $X_{i2}$, ..., $X_{iN}$ drawn from class $\omega_i$ ($N > n$), one obtains the following maximum likelihood estimators for $\mu_i$ and the elements $(\Theta_i)^{uv}$ of the covariance matrix ($1 \leq u, v \leq n$):

$$\hat{\mu}_i = \frac{\sum_{k=1}^{N} X_{ik}}{N}$$

$$\left(\hat{\Theta}_i\right)^{uv} = \frac{\sum_{k=1}^{N} (X_{ik}^u - \hat{\mu}_i^u)(X_{ik}^v - \hat{\mu}_i^v)}{N}$$

The distribution $p(X|\omega_i)$ takes the form

$$p(X|\omega_i) = \frac{1}{(2\pi)^{n/2}\, |\hat{\Theta}_i|^{1/2}}\; exp\; \left(-\frac{1}{2}(X - \hat{\mu}_i)^T \hat{\Theta}_i^{-1}(X - \hat{\mu}_i)\right)$$

Once probability distributions have been obtained for all classes $\omega_i$, one can use Bayes' rule to compute the conditional probabilities that an observation with value $X$ belongs to class $\omega_i$:

$$p(\omega_i|X) = \frac{p(X|\omega_i)\, p\,(\omega_i)}{\Sigma_{k=1}^{M}\, p\,(X|\omega_k)\, p(\omega_k)}$$

Here, $p(\omega_i)$ denotes the a priori probability for the occurrence of class $i$. If those probabilities are not known, one often assumes that membership is equally distributed among the $M$ classes, i.e., $p(\omega_i) = 1/M$. In that case we obtain:

$$p(\omega_i|X) = \frac{p(X|\omega_i)}{\Sigma_{k=1}^{M}\, p\,(X|\omega_k)}$$

Maximum likelihood now simply means that the observation is assigned to the class whose value $p(\omega_i|X)$ is maximal.

### 2.2.3 Validation and Interpretation

Based on the results of the classification, one can now form an initial collection of environmental data objects that are more condensed and therefore easier to interpret than the raw measurements. Then one seeks to identify those environmental data objects that seem interesting or noteworthy according to a predefined set of criteria. This may in particular involve the extraction of data objects that seem to signify *unusual* events or developments.

Because of the uncertainties associated with the data acquisition process, validation procedures need to be employed to avoid undetected hazards, false alerts, and other mishaps [PW95]. One usually starts with *in situ*, context-free validation procedures. Recent measurements are compared to previous measurements *(temporal validation)* and possibly to some reference data obtained under similar circumstances. Measurement values that do not fit the usual patterns are subjected to cross-validation with measurements from other sensors in the same area that measure the same parameter *(geographic validation)*. This may also include previous measurements from those sensors *(space-time validation)*. Those values that still do not fit the norm are forwarded to a cross-validation with sensors that measure different parameters *(inter-parameter validation)*. This last step is highly labor-intensive and requires considerable knowledge about the underlying analytical chemistry, about the land use that is typical for the site in question, the chemical substances that are common there, and so on. As we shall see later, however, any attempt to automate the measurement and validation process is futile if it does not take these various knowledge sources into account.

Obvious measurement errors and outliers are then removed from the output and not taken into account by later processing steps. Whenever one detects continual irregularities in the measurement process, one triggers an operation alert. This alert should eventually result in a closer examination of the sensor in question and, if necessary, appropriate maintenance. It is obvious that this validation procedure is somewhat biased towards avoiding false environmental alerts, rather than making sure no hazard remains undetected. For the latter case, special drills are necessary, where one induces a test substance into the system that is harmless but detectable.

## 2.3 Advanced Techniques

### 2.3.1 Knowledge Representation

For the processing and initial evaluation of environmental raw data, *knowledge-based systems* (or *expert systems*) have considerable potential. Large parts of the data capture process are routine labor; they involve mechanical applications of well-known techniques such as the ones described above. Given the recent progress in the areas of knowledge representation and reasoning, those tasks can be at least partly automated. The key idea is to relieve human experts from the more mundane aspects of the data capture process. Even if full automation is infeasible or undesirable for whatever reason, it is often possible to use a knowledge-based system as a desktop assistant that enables less qualified personnel to perform the evaluation *in most cases*. Only the noteworthy and unusual data items are later forwarded to higher-qualified experts for detailed inspection.

With regard to *knowledge representation*, the requirements of environmental applications can usually be met by standard database and artificial intelligence techniques.

*Static knowledge* is stored in specialized file systems, or in relational or object-oriented databases. An important advantage of object-oriented databases is that they give users the ability to group similar objects into *classes* and to connect those classes in an *inheritance hierarchy*. The objects in each class all share a set of *attributes* and possibly a number of *methods*, i.e., special procedures that usually take one or more objects of the class as arguments. The idea of inheritance is that attributes and methods that are defined for some class $C$ higher up in the inheritance hierarchy are also valid for all classes in the subtree below $C$.

Figure 2.2 gives a simple example. Boxes correspond to classes, and the terms on the right denote concrete objects, i.e., members of these classes. Note that objects may belong to more than one class. In that case, the attributes and methods of all containing classes apply to the object in question.

We delay a further discussion of class hierarchies to Chap. 3, where we discuss relational and object-oriented databases, as well as other storage structures, in great detail.

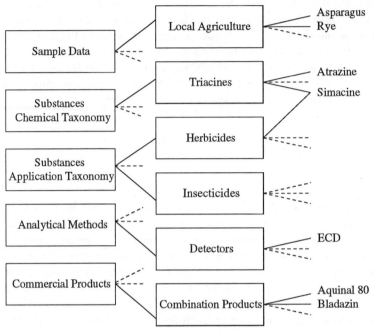

**Fig. 2.2.** Class hierarchy for measurement data evaluation

Besides standard relational and object-oriented databases, IF-THEN rules are among the commonest approaches to representing *dynamic knowledge* in environmental information systems. The idea of a rule-based system is to encode the available knowledge in a possibly large number of relatively simple *rules* rather than hiding it in a complex procedural program. Each rule consists of two parts. The IF part describes a *pattern* that determines when the rule can be applied. The THEN part contains the *action* to be performed in this case. Starting from some initial state, the system checks which of the rules are currently applicable. If more than one rule can be applied, the system picks one of the rules based on a given priority scheme. The action in the THEN part of the rule may just cause a message to be sent to the user. It may also imply a change of the current state because it changes the truth value of some facts from false to true, or vice versa. In that case, other rules may become applicable because their respective patterns are matched. For an example, consider the following two rules from the domain of groundwater measurement.

IF the local agriculture includes corn, grapes, or asparagus
THEN it is likely to find atrazine in the groundwater.

IF a substance has been proven to be in the sample
AND this substance is part of a commercial herbicide $H$
THEN it is likely to find other substances from $H$ in the sample.

The first rule concerns some application-specific knowledge about the habits of the local farmers. The second rule is somewhat different in nature. It refers to "world knowledge" in the artificial intelligence sense. The rule is completely straightforward, and domain experts would use it intuitively, but in a knowledge-based system it needs to be expressed explicitly. The THEN part is somewhat unspecific in both examples. The following section will show how to encode this information in a more rigid framework, and Sect. 2.4 will present a detailed example.

There are numerous variations of the basic rule-based scheme described above, depending on the kinds of action, the priority scheme, and so on. Many expert system shells, as well as the PROLOG programming language, are based on this paradigm. Rules are usually added and updated by means of a graphical user interface. Most commercial expert system shells provide such tools. For more details, we refer the reader to the standard textbook literature in artificial intelligence. Luger and Stubblefield [LS89b], for example, give a good overview.

## 2.3.2 Data Fusion and Uncertain Information

As we saw in the previous section, environmental data capture can be performed with techniques that are more or less standard in the areas of statistical classification, database management, and artificial intelligence. If one is now trying to combine the various pieces of information that are available at the beginning of the capture process, however, one requires more sophisticated concepts. In particular, this *data fusion* process is often based on techniques for the management of *uncertainty*.

When raw imagery and measurement data is aggregated and evaluated, the input data is only one of many sources of information. Other circumstantial information is also taken into account in order to extract those environmental data objects that the user is interested in. Examples of such circumstantial information include:

- experience with the methods and technical devices that are used for data capture
- knowledge about the chemical substances and products involved
- information about the circumstances of the sample (local agriculture, temperature, weather, date, time of day, etc.).

Human experts always take such information into account when evaluating a sample, albeit sometimes unconsciously. Any attempt to automate this process without including such circumstantial information is doomed to failure.

A promising strategy is to form a working hypothesis, and to support or challenge this hypothesis based on the information available. This technique somewhat resembles a lawsuit that is based on circumstantial evidence. The question is how to weigh the different information inputs, and how to combine the resulting evidence for and against the current working hypothesis. This has to include the possibility that the inputs may partly contradict each other.

One option would be to base this strategy on Bayesian probability theory. Each piece of evidence is weighed with a probability between zero and one. While this technique can be used in principle, it has been shown to have several disadvantages in this context. The main problems are:

- There is no way to consider the evidence *pro* and the evidence *contra* independently. Each piece of evidence is weighed with exactly one probability $p$ that denotes the degree of support that this evidence gives to the working hypothesis. In the Bayesian model, however, this also means that it gives support of weight $1 - p$ to the counterhypothesis. This is not always sufficient to model the actual situation, where the degrees of support *pro* and *contra* may be less dependent on each other.
- Bayesian probability theory has been proven to be sensitive to inaccuracies in the input probabilities. In particular, these inaccuracies may propagate through the model in a superlinear manner.
- The Bayesian model often requires that events are independent of each other. This assumption is rarely true in real life.

Recent research in applied mathematics and artificial intelligence has resulted in a number of alternative approaches to handling uncertainty. On the symbolic level, most notably there are *certainty factors* [Ada84] and *fuzzy sets* [Zad65]. More recently, *neural networks* have attracted major interest as a broad class of techniques to represent uncertainty at the subsymbolic level [MP69, RMtPG85]. Keller [Kel95] and Ultsch [Ult95] give several examples of how to apply neural networks to environmental problems.

There is a large number of books that discuss these and other techniques for handling uncertainty in much greater detail. Besides the book by Luger and Stubblefield [LS89b], the reader may consult, for example, the collections edited by Kruse et al. [CKM93, RKV95] and the textbook by Spies [Spi93] (in German).

In this book we discuss one symbolic approach as an example that has been applied successfully to a variety of environmental data capture problems: the *support intervals* of Dempster and Shafer [Dem67, Dem68, Sha76]. The key idea of support intervals is that one should logically separate the arguments *for* and the arguments *against* a given hypothesis, rather than assuming one to be the complement of the other. The Dempster/Shafer theory manages this separation by distinguishing between two concepts: *belief* and *plausibility*. Both concepts are represented by a number between zero and one. The belief $B(H)$ represents the weight of the facts supporting the working hypothesis $H$. The plausibility $Pl(H)$, on the other hand, is one minus the weight of the facts speaking against $H$. If $\overline{H}$ denotes the hypothesis that $H$ is false, we therefore obtain

$$Pl(H) = 1 - B(\overline{H})$$

The belief for the counterhypothesis $B(\overline{H})$ is sometimes referred to as *doubt* with respect to the working hypothesis $H$. We denote this variable as $D(H)$ and obtain

$$Pl(H) = 1 - D(H)$$

In Bayesian probability theory, the concepts of belief and plausibility coincide, i.e.,

$$p(H) = B(H) = Pl(H) = 1 - p(\overline{H})$$

In the Dempster/Shafer framework, on the other hand, the two numbers only have to fulfill the constraint that the belief has to be less than or equal to the plausibility, i.e.,

$$B(H) \leq Pl(H)$$

The difference between $B(H)$ and $Pl(H)$ represents the current degree of uncertainty $U(H)$ about the hypothesis. It is typically one, or close to one, at the beginning of the data fusion process. The more facts are included in the investigation, the smaller becomes the degree of uncertainty. Figure 2.3 gives a graphical representation of these concepts; an example will be given in the following section.

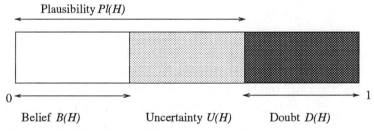

Fig. 2.3. Belief and plausibility of a hypothesis

The feature of this approach that makes it particularly suitable for data capture is its ability to combine different pieces of evidence in a clean and intuitive manner. The basis for this feature is *Dempster's rule*. Suppose there are two facts that both provide some support but also some doubt regarding the hypothesis $H$. Let $B_i(H)$ $(i = 1, 2)$ denote the corresponding beliefs, $Pl_i(H)$ the corresponding plausibilities, and $U_i(H)$ the corresponding degrees of uncertainty. According to Dempster's rule, the belief $B(H)$ that represents the *combined* support is given by the following formula:

$$B(H) = 1/K \cdot (B_1(H)B_2(H) + B_1(H)U_2(H) + U_1(H)B_2(H))$$

where $K = 1 - B_1(H)D_2(H) - D_1(H)B_2(H)$. An analogous rule can be stated to combine the two plausibilities $Pl_1(H)$ and $Pl_2(H)$. Let $D_i(H) = 1 - Pl_i(H)$ denote the two doubts. We then obtain

$$\begin{aligned} Pl(H) &= 1 - D(H) \\ &= 1 - 1/K \cdot (D_1(H)D_2(H) + D_1(H)U_2(H) + U_1(H)D_2(H)) \end{aligned}$$

The new uncertainty is simply

$$\begin{aligned} U(H) &= Pl(H) - B(H) \\ &= U_1(H)U_2(H)/K \end{aligned}$$

It can easily be shown that the new uncertainty is no greater than either of the two input uncertainties, i.e., $U(H) \leq U_1(H)$ and $U(H) \leq U_2(H)$. If $K$ ever turns out to be zero, the two input facts are said to be *categorically incompatible*. In this case the joint beliefs and plausibilities are undefined.

The rationale behind Dempster's rule can best be explained by the combination matrix depicted in Fig. 2.4. The $x$-axis in that matrix represents the first piece of evidence, which provides a belief to support $H$ of $B_1(H)$, an uncertainty $U_1(H)$, and a doubt $D_1(H)$, with

$$B_1(H) + U_1(H) + D_1(H) = 1$$

Analogously, we mark the three entries on the $y$-axis with the belief $B_2(H)$, the uncertainty $U_2(H)$, and the doubt $D_2(H)$, with

$$B_2(H) + U_2(H) + D_2(H) = 1$$

The initial weights for each field in the matrix are obtained by multiplying the weights of its corresponding row and column entries. The problem with this initial weighting, however, is that two of the entries correspond to meaningless combinations: (i) the entry combining $B_1(H)$ with $D_2(H)$, and (ii) the entry combining $B_2(H)$ with $D_1(H)$. In a second stage, these weights have to be set to zero, and the other weights have to be increased proportionally to add up to one again. This can be achieved by dividing all these other weights by $K = 1 - B_1(H)D_2(H) - B_2(H)D_1(H)$. We then obtain the belief $B(H)$ by adding up the modified weights for the entries combining $B_1(H)$ with $B_2(H)$, $B_1(H)$ with $U_2(H)$, and $B_2(H)$ with $U_1(H)$. $D(H)$ is obtained analogously,

|         | $B_1(H)$ | $U_1(H)$ | $D_1(H)$ |
|---------|----------|-----------|----------|
| $B_2(H)$ | Belief   | Belief    | Conflict |
| $U_2(H)$ | Belief   | Uncertainty | Doubt   |
| $D_2(H)$ | Conflict | Doubt     | Doubt    |

Fig. 2.4. Combination matrix to illustrate Dempster's rule

and $U(H)$ comes out as the updated weight of the entry combining $U_1(H)$ with $U_2(H)$.

This combination of evidences can be applied as often as required, and in any order. Compared to Bayesian probability theory, this approach has a number of advantages in many data capture situations. On the one hand, the Dempster/Shafer approach does not require the inputs to be stochastically independent. Of course, it *does* require them to be logically independent – in other words, no piece of evidence is allowed to be weighed more than once.

The problem remains how to obtain the initial values for the belief and the plausibility that a particular fact gives to the hypothesis. As in probability theory, these estimates have to be made by the user, and are usually subject to a meticulous tuning process. It is claimed, however, that the Dempster/Shafer approach is often more robust than probabilities, i.e., small changes in the inputs do not affect the results in a major way [Spi93].

The following section describes a system for the analysis of water measurement data that is based on knowledge-based techniques and the Dempster/Shafer approach.

## 2.4 Water Measurement Data

### 2.4.1 Data Acquisition

Water monitoring is an essential component of any environmental management and protection strategy. The two most important techniques to evaluate a given set of raw sensor data are *chromatography* and *mass spectrometry*. A detailed explanation of the analytical chemistry underlying those techniques is beyond the scope of this book. Here we will only discuss the chromatography approach in some more detail.

Chromatography serves to separate a mixture of substances into its components and to determine these components and their concentration in the original mixture. The basic idea is to warm up the given mixture in a very gradual manner until one sees one component after the other evaporating. The time between the beginning of the process and the evaporation of a substance is called the *retention time*. The evaporating substances are led through a detector that, usually with the aid of a microprocessor, translates the passing of the substance into a peak on an $x$-$y$-diagram, the *chromatogram*. Its $x$-axis represents the retention time, and the $y$-axis the concentration of the corresponding substance.

Characteristic chromatograms are now widely available on the Internet. Figure 2.5, for example, shows a gas chromatogram taken from the Web site of J&W Scientific (http://www.jandw.com). The chromatogram indicates the presence of several pesticides. Atrazine, for example, corresponds to peak no. 4.

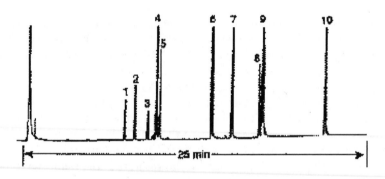

**Fig. 2.5.** Gas chromatogram for pesticide detection [Sci97a]

Chromatography is one of the most important techniques in analytical chemistry. It is very sensitive, yet relatively inexpensive. More advanced techniques such as mass spectrometry are even more specific but are not always economically feasible. In other words, there is a cost/benefit tradeoff, and one has to decide on a case-by-case basis which technique to use for which application. Combined approaches are common as well, starting with a low-cost technique (such as chromatography), and using more expensive approaches (such as mass spectrometry) only for those cases that are doubtful or especially relevant.

The main problem with chromatography data is that the raw data is difficult to interpret. It requires experienced human experts to read the chaotic-looking chromatogram that comes out of a modern automatic chromatographer, and to extract from it a list of components, together with their respective concentrations. In the following section, we present a knowledge-based approach to automate this interpretation process to a certain degree.

## 2.4.2 WANDA: Automating Water Analysis

The acronym WANDA stands for Water ANalysis Data Advisor. From 1989 until 1992, WANDA was a research project at FAW, an applied computer science research institute in Ulm, Germany. It was funded by a consortium consisting of the Environmental Ministry of the State of Baden-Württemberg, Hewlett-Packard Germany, and IBM Germany. The objective was to design and implement a knowledge-based system for the analysis of water measurement data. One of the possible application scenarios was the early detection of pesticides (such as atrazine) in drinking water. The prototype resulting from the project is now being used by the Baden-Württemberg Environmental Protection Agency. For a comprehensive project overview see [Sch93].

The goal of the project was to construct a software system that supports the interpretation of a given gas chromatogram, producing a list of components and their respective concentrations. While the possibility of a fully automatic system was discussed initially, the WANDA team quickly settled for a more realistic solution, where the system is used as an interactive desktop assistant. The idea was to allow a less qualified user (such as a laboratory assistant) to perform the required analysis work nearly as quickly and successfully as an experienced analytical chemist.

The first step of the (semi-)automatic analysis consists of a comparison of the given chromatogram with a reference diagram that indicates the typical peak positions for a number of common substances. The correlation of a given peak with a reference peak can be quantified using the maximum likelihood technique. The substance with the highest value is then chosen as the most likely one.

For an example consider Table 2.1, which represents the initial analysis of some chromatography data coming from a routine screening of Rhine water. The table gives some proposals for the identification of a given peak $p$ with retention time 1602.2. The retention *index* is the standard retention time of the corresponding substance and the given analysis technique. It marks the $x$-coordinate of the substance's peak in the reference chromatogram. For each proposal, a maximum likelihood probability has been computed based on a suitable normal distribution that indicates the probability that the peak $p$ indeed corresponds to that substance.

**Table 2.1.** Proposed identification of a peak with retention time 1602.2

| Reference Proposal | Retention Index | Match Quality |
|---|---|---|
| atrazine | 1709.9 | 0.0014 |
| desethylatrazine | 1614.6 | 0.3704 |
| desisopropylatrazine | 1600.9 | 0.4461 |

Based on the maximum likelihood technique, one would now assume that $p$ represents the substance desisopropylatrazine. Experienced scientists, however, will be able to deduce some more information from the given data. Knowing that desisopropylatrazine is a metabolite of atrazine, and that the mean halfvalue time of atrazine is about 100 days, one would suspect that atrazine could have been found in the water about 100 days ago. This suspicion is supported by the fact that atrazine is a common component of many herbicides, and that it is often applied prophylactically between April and June.

This example shows that a simple match between a given chromatogram and a reference diagram are hardly sufficient to perform a satisfactory analysis. Many more information sources have to be taken into account in order to get near the performance of a human expert. As discussed above, that may include information about the circumstances of the sample, about the substances that are commonly used in local agriculture, knowledge about the chemical composition of trademark herbicides, and so on. Figure 2.6 summarizes the possible knowledge sources for the identification of a given peak.

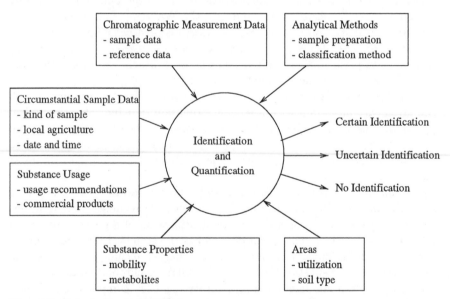

**Fig. 2.6.** Data fusion in WANDA

In WANDA these different sources of information are integrated in a variety of ways. The chromatography data is retrieved directly from the PC attached to the chromatograph. WANDA understands the *Peak96* data format produced by the PC and converts it into some internal representation. Additional sample data, as well as information about the analysis technique, are entered interactively through a graphical, form-based user interface. In a future production version of the system, connections to databases and geographic information systems may replace at least some of this manual input. Figure 2.7 shows the resulting architecture of the WANDA prototype.

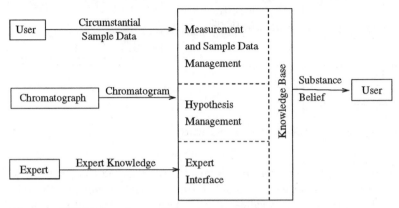

**Fig. 2.7.** WANDA system architecture

In order to weigh and combine those different inputs, the WANDA system uses the Dempster/Shafer technique. Sample data and initial comparisons with reference data are used to create beliefs, while restrictions of the chosen chromatography procedure are the basis for doubts. This approach was chosen to simulate the intuitive procedure a human expert would typically follow. First, one looks at the sample data to see which substances to expect (belief). Then, one performs a reference match for the given chromatogram, which is the basis for further beliefs (Fig. 2.8). The chosen chromatography technique, on the other hand, is usually not efficient for the whole range of possible substances. Some substances may not be detectable at all with the chosen technique, which would be the basis for a (strong) doubt regarding those substances. Note that the reverse is not true: one does not assign a belief based on the fact that a technique *can in principle detect* a substance in the sample.

For a particular peak-substance combination this may look as follows. Initially, there is no particular evidence that a given peak represents, say, atrazine, nor is there evidence against that assumption. This leads us to the initial support interval

atrazine [0.0;1.0]

The reference match may yield a reasonable correlation with a reference peak corresponding to atrazine, which changes the support interval to, for example,

atrazine [0.6;1.0]

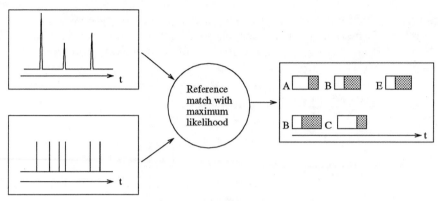

**Fig. 2.8.** Generation of beliefs by reference match

The maximum likelihood probability $p(\omega_i|X)$ corresponding to $\omega_i$ = atrazine is usually a good initial value for the belief. But it is not the only possible one. As noted above, the choice of values is subject to a complex tuning process.

One may subsequently realize that asparagus is cultivated near the location of the sample, which triggers the rule

IF the local agriculture includes corn, grapes, or asparagus,
THEN it is more likely to find atrazine in the groundwater.

This rule increases the belief that the peak in question represents in fact atrazine. Assuming that the belief assigned to the rule is 0.2, one now has to combine the support interval [0.6;1.0] and [0.2;1.0] using Dempster's rule. The new support interval is

atrazine [0.68;1.0]

Finally, one may obtain some information on the analysis technique used, which tells us that the technique can in principle detect atrazine, albeit it is not very reliable for that substance. Depending on the weight assigned to it, this fact could introduce some considerable counterevidence into the scheme. For a weighting of 0.1, i.e., an interval of [0.0;0.1], for example, one obtains a final support interval of

atrazine [0.17;0.25]

We can therefore assume with reasonable confidence that the peak in question does not represent atrazine. Figure 2.9 gives a graphical representation of this gradual deduction process.

Different weights for the counterevidence will be reflected directly in the final result. Consider, for example, a substance that enters a chemical reaction once a certain temperature is reached. If that temperature is below the temperature where the substance leaves the device and passes the detector, that substance cannot in principle be detected by this particular technique. This knowledge could be encoded in a rule such as

Suspicion for atrazine based
on reference chromatogram

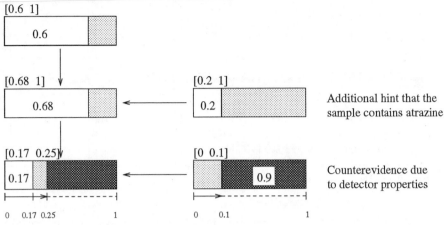

Fig. 2.9. Data fusion using Dempster's rule

IF a substance is thermally unstable,
THEN it cannot be detected by a gas chromatographer.

In the support interval framework, this would mean a strong doubt, which would overrule all beliefs that may have been established previously. If this problem or an equally serious problem had applied to atrazine and the chosen detection technique, one would have had to choose a larger weight to express this fact. An input interval [0.0;0.01], for example, would result in a final support interval of

atrazine [0.02;0.03]

On the other hand, if there had been additional supporting facts instead of counterevidence, then one could have obtained much higher values, such as

atrazine [0.94;1.0].

In WANDA, this stepwise process of establishing evidence is performed for each peak to be identified. On the screen, the process is visualized by maintaining a list for each peak that indicates the substances most likely to match (Fig. 2.10). For each substance one sees the current support interval. The lists are sorted by decreasing belief. As additional evidence is introduced, these lists are updated. In the example, the new evidence caused an order reversal in the lists for peaks 21 and 22. The evidence apparently bolstered the suspicion that peak 21 corresponds to terbutylazine. On the other hand, it became apparent that the method of analysis is not suitable to detect either aziprotryme or desmetryme. As a result, their plausibility values drop to zero. Hence, the procedure now favors sebuthylazine for peak 22. It does not yield any plausible identifications for peaks 23 and 24.

| PEAK NO.: 20<br>RET.TIME: 18.312 | PEAK NO.: 21<br>RET.TIME: 18.844 | PEAK NO.: 22<br>RET.TIME: 19.875 | PEAK NO.: 23<br>RET.TIME: 21.511 | PEAK NO.: 24<br>RET.TIME: 21.697 |
|---|---|---|---|---|
| PROPAZINE<br>Qty[mg]: –<br>Result:[0.79 1.00]<br><br>ATRAZINE<br>Qty[mg]: 0.6701<br>Result:[0.67 1.00] | QUINTOCEME<br>Qty[mg]: –<br>Result:[0.80 1.00]<br><br>TERBUTYLAZINE<br>Qty[mg]: 0.4436<br>Result:[0.79 1.00] | AZIPROTRYME<br>Qty[mg]: –<br>Result:[0.80 1.00]<br><br>SEBUTHYLAZINE<br>Qty[mg]: 8.9838<br>Result:[0.70 1.00] | DESMETRYME<br>Qty[mg]: –<br>Result:[0.64 1.00] | NO SUBSTANCE FOUND |

| PEAK NO.: 20<br>RET.TIME: 18.312 | PEAK NO.: 21<br>RET.TIME: 18.844 | PEAK NO.: 22<br>RET.TIME: 19.875 | PEAK NO.: 23<br>RET.TIME: 21.511 | PEAK NO.: 24<br>RET.TIME: 21.697 |
|---|---|---|---|---|
| PROPAZINE<br>Qty[mg]: –<br>Result:[0.79 1.00]<br><br>ATRAZINE<br>Qty[mg]: 0.6701<br>Result:[0.67 1.00] | TERBUTYLAZINE<br>Qty[mg]: 0.4436<br>Result:[0.85 1.00]<br><br>QUINTOCEME<br>Qty[mg]: –<br>Result:[0.80 1.00] | SEBUTHYLAZINE<br>Qty[mg]: 8.9838<br>Result:[0.70 1.00]<br><br>AZIPROTRYME<br>Qty[mg]: –<br>Result:[0.00 0.00] | DESMETRYME<br>Qty[mg]: –<br>Result:[0.00 0.00] | NO SUBSTANCE FOUND |

**Fig. 2.10.** Peak-substance list before and after the introduction of evidence [Sch93]

In order for WANDA to remain attractive for the user, it is important to keep the knowledge base up to date. New substances have to be added, new or modified analysis techniques have to be taken into account, new rules may have to be created to represent new insights and to correct recognized flaws, misleading items may have to be deleted, and so on. While parts of this maintenance will doubtlessly have to be performed by the vendor of such a system, it also has to be possible for the advanced user, i.e., the analytical chemist, to edit the knowledge base. Such an *expert interface* (cf. Fig. 2.7) is also of crucial importance during the initial customization of the system to suit the particular requirements and habits at the client's laboratory. In this branch of chemistry, the analysis techniques used are usually highly customized and the result of many years of practical experience in each individual laboratory. Detailed adaptation to each particular user is crucial for the commercial success of such a system.

The WANDA prototype was implemented on UNIX workstations using the expert system shell KEE and the programming language Common Lisp from Lucid. The prototype has gone through an extended evaluation at the Baden-Württemberg Environmental Protection Agency in Karlsruhe, Germany. It is used to support the laboratory staff in the screening of Rhine water samples. The typical user mode is a laboratory aide running routine samples through a chromatographer and an attached WANDA program. Based on the WANDA results the aide determines whether everything looks "normal" or whether the sample could be suspicious in some sense. In the latter case, the sample, together with the initial analysis results, is forwarded to a senior member of the laboratory staff for further action.

# 2.5 Evaluating Satellite Imagery

## 2.5.1 Data Sources

Environmental data is increasingly becoming available in digital form from advanced multispectral sensor systems or from scanned aerial photographs. While there has been a notable trend towards using imagery for many years, it has recently gained considerable momentum due to the "peace dividend" mentioned above. Military contractors are trying to recover some of the revenue that has been missing since the end of the cold war and are entering new markets. Satellite technology is one of the technologies that are most easily converted from military to civilian purposes.

There are a variety of satellite systems dedicated mainly to environmental data capture. Table 2.2 gives an overview of some of the best-known systems.

**Table 2.2.** Comparison of satellite systems [Bel95]

| Satellite<br>Sensor | Landsat<br>TM | Landsat<br>MSS | NOAA<br>AVHRR |
|---|---|---|---|
| Main applications | land use | land use | meteorology, land use |
| Spatial resolution | 30 m | 80 m | 1.1–4 km |
| Surface/scene [km] | 185×172 | 185×172 | 2 000×1 000 |
| Repeat coverage [d] | 16 | 16 | 1–8 |
| First launch | 1982 | 1978 | 1986 |
| Number of bands | 7 | 4 | 5 |
| Spectral range [$\mu$m] | 0.45–12.4<br>blue-TIR | 0.50–1.10<br>green-NIR | 0.58–12.50<br>green-TIR |
| Min. recorded scale | 1:75 000 | 1:100 000 | 1:3 000 000 |
| Max. recorded scale | 1:250 000 | 1:500 000 | 1:5 000 000 |
| Price/scene [U.S.$] | 4,500 | 800 | 15 |

| Satellite<br>Sensor | SPOT<br>HRV multispectral | SPOT<br>HRV panchromatic |
|---|---|---|
| Main applications | land | land |
| Spatial resolution | 20 m | 10 m |
| Surface/scene [km] | 60×60 | 60×60 |
| Repeat coverage [d] | 3–26 | 3–26 |
| First launch | 1986 | 1986 |
| Number of bands | 3 | 1 |
| Spectral range [$\mu$m] | 0.50–0.89<br>green–NIR | 0.51–0.73<br>green–red |
| Min. recorded scale | 1:50 000 | 1:25 000 |
| Max. recorded scale | 1:100 000 | 1:100 000 |
| Price/scene [U.S.$] | 3 000 | 4 000 |

The *Landsat* system [Uni97a] provides periodic high resolution multispectral data of the Earth's surface on a global basis. Starting with its inception in 1972 (then called Earth Resources Technology Satellite – ERTS), the program was managed by the U.S. Federal Government. In 1985, there was a politically motivated privatization effort that resulted in the founding of the Earth Observation Satellite Company (EOSAT). The Clinton Administration decided to reverse this decision. Since 1993, Landsat has again been controlled by the U.S. Federal Government, represented by the National Oceanic and Atmospheric Administration (NOAA). The current operational satellite, Landsat 5, was launched in March 1985. Landsat 6 was lost shortly after its launch in October 1993.

Landsat orbits the Earth once every 16 days, taking pictures of each part of the Earth's surface. Each scene covers an area of 185 by 172 square kilometers. One TM scene now costs around $4,500; partial scenes are also available. There are two sensors on board the Landsat system: the Thematic Mapper (TM) and the older Multispectral Scanner (MSS). TM captures data with a resolution of 30 meters and in seven channels (bands) for different parts of the electromagnetic spectrum. Four of the seven bands cover the visual spectrum (0.45–0.8 $\mu$m), the others are infrared (up to 12.4 $\mu$m). MSS has four spectral bands (0.50–1.10 $\mu$m, i.e., green to near infrared) and a resolution of 80 meters. Landsat images are used in a broad range of applications, including land use monitoring, map making, pollution monitoring, snow extent assessments, erosion detection, and forest fire monitoring.

*SPOT (Satellite pour l'observation de la terre)* [Spo97] is an Earth observing satellite coordinated by the French government; partners are Sweden and Belgium. There are three SPOT satellites currently in space, launched in 1986, 1990, and 1993. Each of the three satellites carries two High Resolution Visible (HRV) sensors that can work in two modes. The multispectral mode corresponds to a ground resolution of 20 meters and three channels or spectral bands (0.50–0.89 $\mu$m, i.e., green to near infrared). The panchromatic mode only covers the spectrum between 0.51 and 0.75 $\mu$m (green to red) but obtains a 10 meter ground resolution. SPOT features some sophisticated scanning technology to support stereoscopic imagery and other advanced viewing options. Main applications include environmental impact studies, geologic exploration, and thematic map making. Since 1997, the SPOT Web site also allows access to the DALI image archive, administered by the Centre National d'Etudes Spatiales. This browsable archive contains more than 4 500 000 SPOT images dating back to 1992. Given a location's coordinates, the server returns up to five images of the requested area, in either color or black and white, with less than 10% cloud coverage.

*AVHRR*, the *Advanced Very High Resolution Radiometer* [Uni97b, LO93] is another satellite system managed by the NOAA. Other than Landsat, however, AVHRR is mainly geared towards meteorological and environmental applications. Typical application areas are weather forecasting, pollution monitoring, toxic algal bloom detection, and sandstorm monitoring. There are two AVHRR satellites, one with four and one with five spectral bands. One band is in the visual range ($V$), one in the near infrared ($NIR$), the other two or three in the middle and far infrared. The quotient $(NIR-V)/(NIR+V)$

has been found empirically to be a good indicator of vegetation vigor. Under the name *Normalized Difference Vegetation Index (NDVI)* it is used widely to map global vegatation on a regular basis [LK87]. AVHRR ground resolution is only about 1 km. On the other hand, the two AVHRR satellites provide two images per day of any given area, i.e., daily global coverage.

The U.S. Geological Survey's EROS Data Center (EDC) has begun to develop AVHRR time series data sets [LMOB91, LO93]. EDC produces four levels of data products that are designed for both the biophysical and land cover data requirements of global change researchers:

- processed daily AVHRR scenes of most of North America and selected other areas
- composite images for specified time intervals for the conterminous U.S., Alaska, Mexico, and Eurasia
- time series sets for those areas
- a land cover characteristics prototype database for the conterminous U.S.

Aerial and satellite imagery is increasingly available through the World Wide Web. NASA, for example, is preparing a Web version of its Earth Observing System Data and Information System (EOSDIS). [Uni98b, DPSB97] EOSDIS contains data from all satellite systems described above and others. The company Earth Observation Sciences (EOS) is running a Web-based Catalogue and Browse System (CBS) to provide instant access to aerial and satellite imagery (http://www.eos.co.uk). CBS allows users to search for data both temporally and spatially, possibly using additional parameters defined by the data archive (such as cloud cover or orbit number). The search functionality includes a global map interface with user-defined polygon searching and complex temporal searches such as searching for annual repeating time periods. CBS provides the ability to browse the results returned from a query before downloading, requesting or ordering data.

## 2.5.2 Data Processing

The digital raster data obtained from those satellite systems contains the desired environmental information in an implicit form only. Moreover, the same areas and phenomena are never recorded the same way because of the ever changing conditions of the atmosphere, the illumination, and the phenology of vegetation. Analyzing this kind of imagery is a well-developed branch of engineering. It involves techniques from a variety of disciplines, including signal processing, statistics and classification, and, increasingly, artificial intelligence. Problem areas include the classification of a given pixel, the grouping of adjacent pixels to form objects, and the recognition of features. Similar to the structure presented in Sect. 2.2, one can distinguish three phases: iconic image processing, classification, and symbolic image processing [GHM+93]:

**Iconic Image Processing.** Iconic image processing essentially corresponds to the *raw data processing* described in Sect. 2.2.1. Here one applies a variety of image processing algorithms to the incoming raw data in order to produce an appropriate visualization. This typically includes [SD78]:

- rectifications of the image geometry to match the scale and coordinates of applicable maps
- corrections of the spectral intensities by contrast enhancement;
- elimination of sensor faults
- noise suppression
- edge improvement.

In order to emphasize certain features, one may combine selected spectral bands, often by simple arithmetic operations, and assign the resulting raster data to the RGB (red/green/blue) video signal to display a color image. The computation and visualization of the NDVI vegetation index, as described above, is a typical example.

In general, this first processing step is heuristic in nature and based on human experience and expertise. The overall goal is to produce a set of images that are visually appealing and easy to interpret with respect to the given task.

**Classification.** This second step serves to classify the imagery with respect to a given, application-specific taxonomy (such as land use). Here one usually applies techniques like the ones presented in Sect. 2.2.2. Features known to be typical for the class to be recognized are propagated over the whole image to find all regions with similar characteristics. The training areas, which are typically selected manually, determine the expected spectral reflectance of those classes. For each class $\omega_i$, one computes the average signatures $\hat{\mu}_i$ of all pixels $X_{ik}$ in all corresponding training areas. Remember that

$$\hat{\mu}_i = \frac{\Sigma_{k=1}^{N} X_{ik}}{N}$$

is a point in multidimensional feature space. The dimensions of this space are not restricted to the channels of the incoming signal but may also be used to encode secondary information. A pixel to be classified belongs to a class if its signature does not differ too much from the average signature of the corresponding training areas; see Sect. 2.2.2 for more details.

The procedure works well if the signatures of the different classes are sufficiently homogeneous and do not overlap in feature space. If that is not the case, the average signature may not be sufficient to characterize a given class, and one must take other criteria into account. Some classes may also be characterized by combinations of signatures. For an orchard, for example, one typically obtains a variety of spectral values that are mixtures of the characteristic signatures for trees and for grass. Other classes may be recognized based on their texture rather than on their spectral signature [Zam88].

Once again, this step usually involves a considerable amount of human interaction. It is a heuristic process based on experience, technical expertise, and a thorough knowledge of the application domain and the requirements of those who use the imagery data for their decision making.

**Symbolic Image Processing.** Symbolic image processing corresponds to the *validation and interpretation* phase described in Sect. 2.2.3. The data on the resulting image is condensed, grouped into objects, interpreted, and associated with geographic entities. Usually this involves the conversion of the incoming raster data to a vector representation that is more compact and easier to process [EGSS91, Rie93]. The geographic entities may be administrative units such as states, districts, counties, or land parcels. They may also be ecological units such as forests, agricultural areas, or residential areas.

For this processing step it is usually indispensable to interact with a GIS that contains data about the regions in question [EGSS91, Dav91]. This input data can help to improve the classification results, for example, by providing a priori probabilities (cf. Sect. 2.2.2). Symbolic image processing once again requires experience and knowledge that is not contained in the image itself. The data sets resulting from this step are often ready to be stored in a GIS again although problems of accuracy and format compatibility have to be taken into account [LCF+91].

Figure 2.11 illustrates the three processing steps described above. Figure 2.11a shows the original Landsat TM data. It is taken from the near infrared channel, which is appropriate to discriminate land and water-covered areas. Figure 2.11b shows a water mask generated from this data by a threshold classification. Figure 2.11c shows water objects derived from the water mask by vectorization and some cartographic generalization.

a satellite imagery         b threshold classification         c water objects

**Fig. 2.11.** Generation of water objects from Landsat TM imagery [GHM+93, with kind permission from Kluwer Academic Publishers]

Most of the operations in this sequence are *unreliable* in the sense that they introduce uncertainties into the classification result. Error sources include atmospheric conditions, sensor operation, geometric inaccuracies, and human judgement. Many of these errors cannot be avoided completely. It is therefore necessary to monitor possible error sources and to assess their impact on the result of the image classification. This usually involves a cost/benefit analysis. Lunetta et al. [LCF+91] give a comprehensive overview of error sources and possible remedies.

If one obtains processed remote sensing data without complete or reliable information about its history (the *lineage*), it may be desirable to verify the classification results by means of sampling. Moisen et al. [MEC94] give a good comparative overview of sampling methods that are common in remote sensing, including simple random sampling, systematic sampling, and cluster sampling. For an introduction to the underlying statistical concepts, see the textbook by Cochran [Coc77].

For many years, researchers have looked for suitable knowledge-based techniques to support the interpretation of remote sensing data [Goo87, McK87, Des90, TSJ91]. In the following section, we describe one such attempt in some more detail.

### 2.5.3 RESEDA: A Knowledge-Based Approach

The goal of the research project *RESEDA (REmote SEnsor Data Advisor)* was to design and implement a knowledge-based system for the analysis of satellite imagery and extraction of environmental information. The project was conducted between 1989 and 1993 at FAW Ulm. It was funded by the Environmental Ministry of the State of Baden-Württemberg and Siemens Nixdorf Informationssysteme AG.

In the subsequent presentation we follow the article by Günther et al. [GHM+93]. For a complete overview of the project see the book edited by Günther and Riekert [GR92] (in German). The RESEDA system architecture is based on three key components (Fig. 2.12):

1. An *image processing system* supports the basic iconic image processing operations described above.
2. A PROLOG-based *expert system* manages models for the derivation of environmental information from the image raw data and administers the knowledge required to apply those models.
3. A *geographic information system* manages the geographic information used to support the image interpretation process. Results of the evaluation may be used in turn to enhance or extend the spatial data stored in the GIS.

For the implementation of the prototype the RESEDA team concentrated on Landsat TM imagery. The analysis is restricted to three of the seven TM

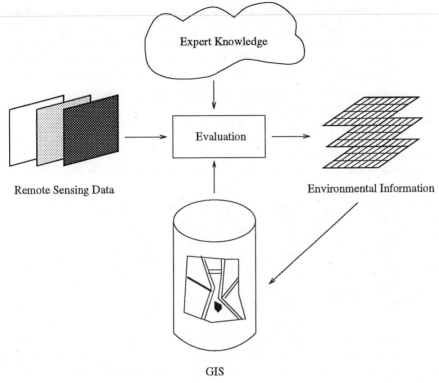

**Fig. 2.12.** Information flow in RESEDA

spectral channels: TM3, TM4, and TM5. Each pixel can thus be represented by a three-dimensional point in feature space. Pixels are classified with respect to one of the following eight land use classes:

1. *agricultural*
2. *greenland*
3. *orchard*
4. *vineyard*
5. *forest*
6. *urban*
7. *rock*
8. *water.*

For each class one needs a training area to establish a three-dimensional normal distribution for pixels belonging to that class. As outlined in Sect. 2.2.2, under certain simplifying assumptions the probability that a pixel with value $X$ belongs to class $\omega_i (i = 1...M)$ is

$$p(\omega_i|X) = \frac{p(X|\omega_i)}{\Sigma_{k=1}^{M} p\left(X|\omega_k\right)}$$

These probabilities serve as input to a maximum likelihood analysis that determines the most likely class for each pixel $X$. In order to evaluate the quality of the classification, one selects a sample of pixels whose class membership is known *(ground truth)*. We then check which of those *control areas* are classified correctly based on the spectral properties. The results of this test can be represented in a two-dimensional *confusion matrix*.

Table 2.3 gives an example. For simplicity, we only consider four kinds of land use. The four columns represent control areas for the land uses *agricultural* (1), *greenlands* (2), *orchard* (3), and *vineyard* (4). The bold entries represent the maximum likelihood classification. Note that the *greenlands* and the *vineyard* areas are misclassified as *orchard* and *greenlands*, respectively.

**Table 2.3.** Confusion matrix based on the spectral properties alone

|                        | area 1     | area 2     | area 3     | area 4     |
|------------------------|-----------|-----------|-----------|-----------|
| 1 *agricultural*       | **56.17** | 14.76     | 34.60     | 28.58     |
| 2 *greenlands*         | 16.64     | 21.67     | 16.74     | **32.28** |
| 3 *orchard*            | 7.97      | **32.88** | **38.42** | 10.83     |
| 4 *vineyard*           | 3.65      | 5.07      | 5.12      | 26.09     |
| 5 undetermined/other   | 15.57     | 25.62     | 5.12      | 2.22      |

As this example shows, it may be difficult to obtain a satisfactory classification quality if no circumstantial knowledge about the observed objects is included. One reason is that the different object classes may be so similar with respect to their spectral properties that they are hard to distinguish on the image. Another reason may be that the imagery is noisy, often due to some atmospheric phenomena. Finally, there is also the problem of mixed pixels that occur because the area corresponding to the pixel is in itself heterogenous; we already mentioned an orchard as the typical example. One therefore tries to include secondary knowledge about the observed area into the analysis. Possible sources for such secondary information include data about the atmosphere and the climate, as well as geodata, such as thematic maps or digital terrain models. In RESEDA, this kind of data is extracted from the GIS that is part of the integrated architecture.

The secondary knowledge is integrated with the results of the maximum likelihood analysis using the Dempster/Shafer approach. The likelihoods represent probabilities, and there is no uncertainty in the maximum likelihood scheme. We therefore interpret each likelihood $p$ as a belief *and* as a plausibility, i.e., $p(\omega_i|X) = B(\omega_i|X) = Pl(\omega_i|X)$. These Dempster/Shafer evidences can then be integrated with further evidences based on secondary knowledge. Figure 2.13 gives an overview of possible knowledge sources.

Table 2.4 gives the modified confusion matrix after the introduction of additional evidence. As the comparison with Table 2.3 shows, the percentage of unidentified pixels decreases and the overall quality of the classification increases. Note that the additional evidences are particularly efficient for those classes whose spectral confusion is already good. Both the *greenlands* and the *vineyard* areas are now classified correctly. On the other hand, the *orchard* area is misclassified as *agricultural*. This indicates a lack of evidence for *orchard* in comparison to the support in favor of *agricultural*.

**Table 2.4.** Confusion matrix after introduction of additional evidence

|                        | area 1 | area 2 | area 3 | area 4 |
|------------------------|--------|--------|--------|--------|
| 1 *agricultural*       | **83.17** | 8.37   | **49.49** | 35.17  |
| 2 *greenlands*         | 11.41  | **71.73** | 10.79  | 15.06  |
| 3 *orchard*            | 0.71   | 2.91   | 30.70  | 10.69  |
| 4 *vineyard*           | 1.87   | 1.63   | 5.95   | **37.46** |
| 5 undetermined/other   | 2.84   | 15.36  | 3.07   | 1.62   |

Figure 2.13 gives a basic outline of the data fusion in RESEDA. It is obvious that the various steps of this complex process are difficult to perform by somebody whose expertise in satellite imagery interpretation is limited. Administrators in a typical local environmental protection agency, for example, would rarely be able to go through this kind of analysis themselves. This is exactly where the RESEDA system is able to provide the required technical support.

The RESEDA system serves as an advisory system for the non-expert user to plan and execute the various steps of the analysis. The user just specifies the available data and desired information. From these specifications, the RESEDA system generates a set of processing plans. After one of the plans is selected, it is translated into a UNIX shell script and executed automatically by a plan interpreter. In this context, automatic execution means that the appropriate procedures are called with correct arguments in a correct sequence. Note, however, that those procedures may run again in an interactive mode and require user intervention (for example, to digitize training areas for a supervised classification).

Figure 2.14 gives an example of plan generation and execution. First, users state that they are interested in a vegetation image of some area defined previously. The system proposes that the input data from the TM channels 3, 4, and 5 be processed by means of a special algorithm called *histogram equalization*, and the results combined into an RGB image. Based on this plan, the RESEDA system generates a UNIX shell script that performs the task. Figure 2.15 illustrates the result.

## 2.6 Summary

In this chapter we first presented possible taxonomies for environmental data objects (Sect. 2.1). In Sect. 2.2 we gave a detailed description of data capture, i.e., the mapping of environmental objects to environmental data objects. We distinguished between domain- and device-specific raw data processing and the subsequent classification, which relies on standard statistical techniques. A final validation and interpretation step serves to condense the raw data into a more compact format that can be stored in a database or a GIS.

Section 2.3 discusses several advanced techniques for data capture, including knowledge representation and uncertainty management.

The chapter concludes with two case studies: WANDA (Sect. 2.4) is concerned with water analysis, and RESEDA (Sect. 2.5) is a knowledge-based system for satellite imagery interpretation. Table 2.5 compares the basic features of data capture in WANDA and RESEDA. There are no principal differences between those two applications. In both cases, one first applies maximum likelihood to obtain an initial classification. In a second stage, one introduces additional domain-specific evidence to improve the classification results.

**Table 2.5.** Data capture in WANDA and RESEDA

|  | WANDA | RESEDA |
|---|---|---|
| Raw Data | ground water | satellite imagery |
| Reference Match | comparison of sample peaks with peaks from a reference diagram | spectral comparison of sample pixels with reference classes |
| Reference Data | 1-dimensional normal distribution per reference peak | $n$-dimensional normal distribution per reference class |
| Ranking | maximum likelihood | maximum likelihood |

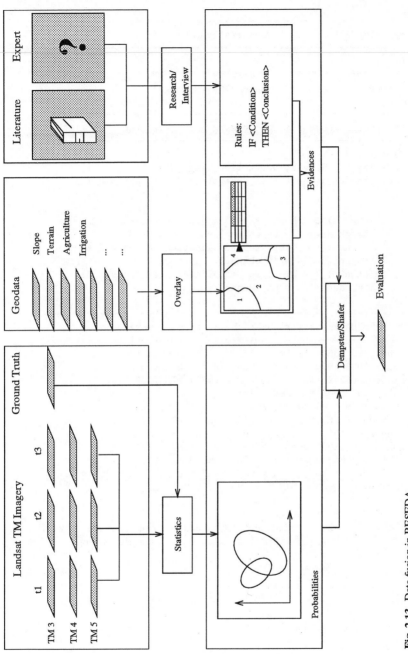

**Fig. 2.13.** Data fusion in RESEDA

```
------------------------------------------------------------------
*** RESEDA Assistant: consultation started
------------------------------------------------------------------
Query      :
[data_type              = display,
 theme                  = vegetation_image]


------------------------------------------------------------------
*** RESEDA Assistant: processing plan generated
------------------------------------------------------------------
Plan No. 1:
 display_9 = display(vegetation_image_8)
  vegetation_image_8 = tm_vegetation_image(tm_channel_3,
                                           tm_channel_5,
                                           tm_channel_7)
   tm_channel_3 = histogram_equalization(tm_channel_2)
     \tm_channel_2 = tm_channel_selection
   tm_channel_5 = histogram_equalization(tm_channel_4)
     \tm_channel_4 = tm_channel_selection
   tm_channel_7 = histogram_equalization(tm_channel_6)
     \tm_channel_6 = tm_channel_selection


------------------------------------------------------------------
*** RESEDA Assistant: UNIX shell script generated
------------------------------------------------------------------
echo 1000 1000 > z_tmF_1.idx
count 1000 1000 none < t5ka4.bld > t5ka4.stat
histequ t5ka4.stat t5ka4.bld z_tmF_1.bld 1000 1000 0
echo 1000 1000 > z_tmF_2.idx
count 1000 1000 none < t5ka5.bld > t5ka5.stat
histequ t5ka5.stat t5ka5.bld z_tmF_2.bld 1000 1000 0
echo 1000 1000 > z_tmF_3.idx
count 1000 1000 none < t5ka3.bld > t5ka3.stat
histequ t5ka3.stat t5ka3.bld z_tmF_3.bld 1000 1000 0
T_pseudo 1 z_tmF_1.bld z_tmF_2.bld z_tmF_3.bld z_vgb_4.bld 2
T_display z_vgb_4.bld 0 0 1000 1000 8cyc.LUT
echo DONE
```

**Fig. 2.14.** Processing plan and UNIX shell script generated by RESEDA [GHM+93, with kind permission from Kluwer Academic Publishers]

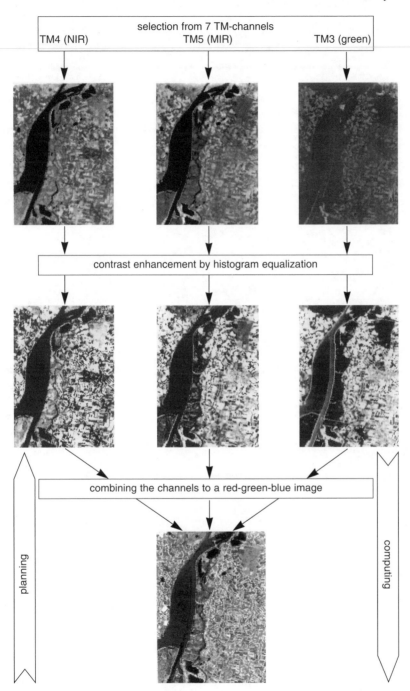

**Fig. 2.15.** Combining data from three TM channels to a vegetation image [GHM+93, with kind permission from Kluwer Academic Publishers]

# 3. Data Storage

The way environmental data is stored is currently changing at a rapid pace. A considerable fraction of relevant data is still only available in analog form. This concerns historical data but also a large number of more recent thematic maps, images, and documents. While this is a problem in environmental information management today, it is not going to be a major impediment in the near future. Those historical data sets that are of relevance in current and future applications are rapidly being digitized. This process is supported by the continuous progress in scanning technology as well as by the emergence of a market for digitizing historical data that is dominated by a handful of specialized software companies and publishers. *New* data is almost exclusively captured in some digital format, and it is mainly a question of logistics to make those digital versions available. The World Wide Web will help to speed up this process considerably.

In this book we concentrate on software techniques for the management of *digital* environmental information. There are essentially two options for storing a given digital data set:

- a *database management system (DBMS)* with a well-defined data model, typically relational, object-relational, or object-oriented
- an application-specific *file system*, as it is still used by many geographic information systems (GIS).

It is important to view this choice independently of the kind of software tools that are used later to query and update the data. While in the past, storage and processing were tightly coupled, more recent systems make a clear distinction between those tasks. This trend is a direct result of the general tendency towards *open systems*. As users demand comfortable interfaces between different hardware and software tools across heterogenous platforms, vendors have been forced to decompose their products along the lines of more narrowly defined functionalities.

GIS in particular used to be monolithic systems, taking care of storage, querying, and visualization of geographic information in a tightly integrated manner. More recently, GIS products tend to follow a toolbox approach instead. Users are increasingly free to choose between a vendor-specific file system and a commercial database system for *storing* the geographic data.

On top of the chosen storage system, GIS then provide advanced function-alities to *query, modify,* and *visualize* the data. Other software tools may be added to compute statistical aggregates, to perform fast full-text searches, and so on.

In this chapter, we first discuss the strengths and weaknesses of com-mercial software systems for the management of very large environmental data sets. Section 3.1 starts with a presentation of geographic information systems and a discussion of the major problem spots. Section 3.2 continues with an analysis how conventional database technology supports the spe-cial needs of geographic and environmental data management. We then show how some recent developments in database research can be utilized for more efficient data management in this domain. In particular, we discuss the de-sign and implementation of spatial data types and operators, abstract data types, and spatial query languages. Section 3.3 presents the state of the art in spatial access methods. Section 3.4 discusses the basic features of object-oriented databases and their potential for the management of environmental data. We also present a case study: GODOT is an object-oriented geographic information system that has been developed on top of a commercial object-oriented database system. Section 3.5 concludes with an overview of likely future developments.

## 3.1 Data Storage in GIS

Geographic information systems are an essential tool for the management and visualization of large amounts of environmental information. Motivated by the potential of computers for cartographic tasks, research and development in this area started about 25 years ago. In the late 1970s, the first systems entered the market. Most of the early customers were from the public sector. Later on, GIS were also used in the context of management information systems and environmental information systems.

While GIS are a crucial component of environmental information man-agement, they are not the main focus of this book. In the following we will assume some basic knowledge of GIS and the underlying concepts. Readers not familiar with GIS are advised to consult the related literature. Com-mon references include a two-volume treatment of the subject by Maguire et al. [MGR91], a collection of introductory readings edited by Peuquet and Marble [PM90b], as well as textbooks by Aronoff [Aro89], Bill and Fritsch [BF97, Bil96] (in German), Bonham-Carter [BC94], Burrough [Bur86], Star and Estes [SE90], and Tomlin [Tom90]. For surveys on database issues in GIS, see the article by Bauzer Medeiros and Pires [MP94] and the collection edited by Adam and Gangopadhyay [AG97]. The relationship between GIS and environmental information systems is discussed in a survey article by Bill [Bil95].

### 3.1.1 Traditional GIS Functionalities

The original idea for GIS was to "computerize" the metaphor of a thematic map. As a result, GIS are designed to support the whole life cycle of map data. More generally, "GIS are computer-based tools to capture, manipulate, process, and display spatial or geo-referenced data" [Fed93]. This definition should remind the reader of the data flow that we observed for environmental data in Chap. 1. Geographic data is a special and important type of environmental information, and the data flow described by Fedra can indeed be viewed as a special case of our scheme, although the details are somewhat different at each stage.

With regard to *data capture*, GIS provide software support for digitizing analog map data. This software is typically used in combination with digitizing tables that are equipped with mouse-like devices for tracing shapes on a given analog map. The movements are captured by the system, and whenever the user clicks on the digitizing device, its current location is translated into coordinates and stored in a file. Modern scanning technology has helped to facilitate this somewhat cumbersome task, but so far it has not been possible to automate the process completely. Maps are an important part of our culture, and as a result we have become used to very high standards for accuracy and presentation. Human understanding of maps is still essential to perform the conversion process from analog to digital in such a way that those high standards are met by digital maps as well.

For *data storage*, GIS used to rely exclusively on customized file system solutions. Since the mid-1980s, vendors have started to use commercial relational database systems. For the alphanumeric (non-spatial) part of the GIS data, these have quickly become the state of the art. The spatial data, however, is still mostly held in proprietary file systems. Most of the underlying data models are *layer-based*: information is encoded in a number of thematic maps, such as vegetation maps, soil maps, or topographic maps. With regard to geometry, each such map corresponds to a partition of the universe into disjoint polygons. Each polygon represents a region that is sufficiently homogenous with respect to the theme of the map. Maps may be enhanced by lines and points to represent specific features, such as roads or cities.

*Data analysis* then mainly consists in intersecting different map layers *(map overlay)* and aggregating the available data in a task-specific manner. Map overlay allows users to connect and interrelate different contents, and visualize the results in new, customized maps. The non-spatial data corresponding to the two input maps needs to be synthesized in a suitable manner, e.g., by creating a new relation whose columns are the union of the columns of the two input relations. Figure 3.1 illustrates this idea on a simple example. To perform aggregations on the non-spatial data, one may use the aggregate functions of the database query language SQL. Additional software modules may be available for more complex statistical aggregation, environmental modeling, and other special functionalities; see Sect. 4.3 for an extensive

discussion. Support for *spatial* aggregation, on the other hand is still rudimentary. The automatic aggregation of cartographic data has proven to be
extremely difficult [McM87, McM88], and once again, manual intervention is
often required.

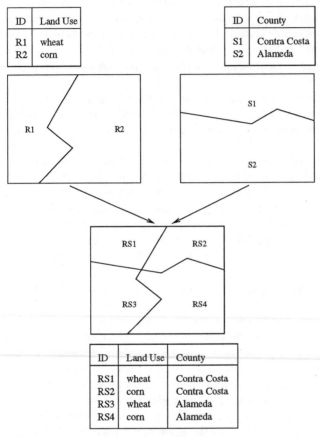

| ID | Land Use |
|----|----------|
| R1 | wheat |
| R2 | corn |

| ID | County |
|----|--------|
| S1 | Contra Costa |
| S2 | Alameda |

| ID | Land Use | County |
|-----|----------|--------------|
| RS1 | wheat | Contra Costa |
| RS2 | corn | Contra Costa |
| RS3 | wheat | Alameda |
| RS4 | corn | Alameda |

**Fig. 3.1.** Overlay of two thematic maps

### 3.1.2 Problem Areas

Most commercial GIS are the result of evolutionary application-oriented developments based on early research activities of the geoscientific community.
Database research has not focused on GIS-relevant questions until fairly recently. By the late 1980s, several conceptual deficiencies in the resulting GIS
designs had become obvious, especially in comparison to modern relational
database systems. Many of these problems can be traced back to their origi-

nal orientation towards file systems, which provide only rudimentary database functionalities and do not scale up very well to large amounts of data.

**Ad Hoc Query Facility.** There are typically three types of interface to a relational database system: a programming language interface; an interactive interface that allows the user to state SQL queries in an ad hoc manner; and a graphical interface, whose expressive power is usually limited compared to the other two options. Many older GIS do not provide an equivalent to the interactive SQL interface. Whenever users want to retrieve or update a data item, they have to go through the graphical interface provided by the system. Most of these interfaces are menu-driven and do not have nearly the expressive power that an advanced database query language such as SQL can offer.

**Persistence.** An important aspect of databases is that they provide persistent storage, i.e., the stored data is available permanently to be used by different applications and ad hoc queries. For that purpose, a database system has to store not only some isolated pieces of raw data but also some structural information to aid the interpretation of the data records. In relational databases, for example, the system maintains catalog information about the different tables in the database in a special relation (the *relation* relation). GIS traditionally store their data on some proprietary file system, which is kept in persistent storage. The problem is that a lot of the related structural information is hidden inside different GIS *programs* that are not accessible to most users. Any attempts to use the data without those programs and without a full understanding of the underlying structure may lead to serious misinterpretations.

**Concurrency.** Because they operate in a multi-user environment, database systems provide mechanisms to facilitate the controlled sharing of data and resources. Certain applications may require that many users be able to read or update a subset of the database concurrently. At other times, one has to make sure that certain data items can only be read or updated by a selected group of individuals. Current database technology provides an array of sophisticated techniques for the management of complex transactions, concurrency, and security. Because of the complexity of these techniques and because of the significant overhead involved, most commercial GIS have originally been designed as single-user systems. Issues of security or concurrency have not been addressed until recently.

**Recovery.** Simple file systems, as used by many GIS, do not provide any facilities to ensure the consistency of the data after system crashes. If a system crash occurs after a user has manipulated certain data items in main memory, it is not immediately clear which updates have been written back to permanent memory before the crash, and which ones have not. This behavior of the system can lead to inconsistent states of the data. In database systems, on the other hand, it is ensured that a transaction (i.e., a sequence of steps that is just one operation from the user's point of view) is carried out either completely or not at all. If a system crash occurs, the system reestablishes a consistent state after coming back up. Ideally, the user does not even notice that some kind of recovery operation has occurred.

**Distribution.** In a distributed database, data can be stored at different locations without jeopardizing the integrity of the database. The distribution of the data is usually transparent to the user, and updates and retrievals can be performed exactly as in the non-distributed case. It is now possible, however, to store data where it is needed most in order to improve system performance, and to establish decentralized ownership privileges without affecting the logical consistency of the database. Distributed data management has made significant progress since the introduction of the relational data model, which greatly facilitates data distribution. This is also true for many object-oriented data models. Traditional GIS data models, on the other hand, do not lend themselves easily to distributed data management, and many GIS therefore do not provide any such features.

**Data Modeling.** Commercial GIS are often too static and inflexible with regard to new applications that may require customized data types and interfaces. Further problems include the modeling of structured objects, and the integration of efficient access methods for very large spatial databases.

**Semantic Integrity.** Most GIS do not offer functionalities to preserve semantic integrity. For example, it is not possible for a user to specify that a value must be included in a particular value range or that it is valid only in connection with certain other values. Modern DBMS, on the other hand, offer triggers and similar techniques to maintain consistency according to user-defined semantic constraints [Sto90, LLPS91].

### 3.1.3 Open GIS: GIS in a Client-Server Environment

Many of these problems of commercial GIS are now disappearing. The reason is that more and more vendors give up on their traditional strategy of selling closed, proprietary systems, in favor of an open toolbox or "open GIS" approach [VS94, VS98].

In 1994, an international group of GIS users and vendors founded the Open GIS Consortium (OGC), which has quickly become a powerful interest group to promote open systems approaches to geoprocessing [OGC97]. The OGC defines itself as a "membership organization dedicated to open systems approaches to geoprocessing." It promotes an *Open Geodata Interoperability Specification (OGIS)*, which is a computing framework and software specification to support interoperability in the distributed management of geographic data. OGC seeks to make geographic data and geoprocessing an integral part of enterprise information systems. Possible applications in the context of environmental information management have been discussed by Gardels [Gar97].

More information about OGC including all of their technical documents are available at the consortium's Web site, http://www.opengis.org.

Due to strong customer pressure, the trend towards such open GIS has increased significantly ever since. Commercial database systems can be integrated into open architectures in a relatively simple manner. A GIS can thus gain directly from the traditional strengths of a modern database system, including an SQL query facility, persistence, transaction management, and distribution. Most GIS vendors have recognized these advantages and offer interfaces to one or more commercial database systems. Moreover, it is often possible to purchase the chosen DBMS bundled together with the GIS at a special price. Although object-oriented databases would often offer superior functionalities, most interfaces are to relational systems, due to customer demand.

Until recently, the database systems were used only for the *non-spatial* data, while the spatial data remained in proprietary file systems. Previous versions of ESRI's GIS ARC/INFO [Mor85, PM90a, ESR98] were a typical example of this approach. For the non-spatial data, the user could choose between a commercial relational database system or ESRI's file system solution (INFO). The spatial data was always stored under ESRI's proprietary system (ARC).

In those cases where the *spatial* data is stored in a relational database system as well, the database usually just provides *long fields* (also called *binary large objects (BLOBs)* or *memo fields*, cf. Sect. 3.2.3) that serve as containers for the geometric data structures. Those are in turn encoded in a proprietary spatial data format. The database system can not interpret the content; it can therefore not provide any data-specific support at the indexing and query optimization level. What the user gains, however, is increased data security and concurrency, as well as accessibility of the spatial data via the ad hoc query interface.

A typical representative of this approach is Siemens Nixdorf's GIS SICAD/open [Sie98]. For data storage, SICAD/open offers a component called GDB-X [Lad97] that provides an interface to several commercial relational database systems (currently Oracle or Informix). As in previous SICAD versions, both spatial and non-spatial data are stored in the same database in an integrated manner. For the spatial data, the relations serve as containers that manage the geometries as unstructured BLOBs.

More recent versions of ARC/INFO also follow this paradigm. ESRI's Spatial Database Engine (SDE) provides the means to store and manage spatial data within a commercial relational database. Users can currently chose between Informix, Oracle, Sybase, and IBM's DB2 [ESR97c].

Yet another direction is pursued by the GIS vendor Smallworld Systems. Their object-oriented Smallworld GIS stores both spatial and non-spatial data in a proprietary data manager called Version Managed Data Store (VMDS) [BN97]. VMDS provides efficient support for version management and long transactions – areas where commercial relational databases are still notoriously weak. External applications can access VMDS data either via an SQL server or via an application programming interface (API). Vice versa, it is possible to access commercial relational databases from Smallworld GIS. When creating a new table, users can choose whether to store it in VMDS or in some commercial DBMS linked to the GIS. Both kinds of tables can be accessed the same way, using the object-oriented concept of overloading (see Sect. 3.4.5). The main difference is that updates to the relational database are visible immediately to all users, whereas updates made to VMDS are only visible in the version in which they were made.

The ultimate open GIS would be a collection of specialized services in an electronic marketplace [VS94, Abe97, VS98]. It is not clear at this point whether this is really what the customer wants. Advantages include vendor independence and scalable cost structures. In theory, users just buy the services they need without any unnecessary extras. On the other hand, this kind of toolbox approach requires considerable expertise on the users' part to select and assemble the right services for their purposes. In a sense, such an approach would be a reversal from the "turnkey" philosophy that says that users should have the full functionality of a system available immediately after installation [The97].

## 3.2 Spatial Database Systems

For storing spatial data efficiently, database systems still have to overcome some of their notorious weaknesses. Spatial database systems represent an attempt by the database community to provide suitable data management tools to developers and users in application areas that deal with spatial data. Note that we consider a spatial database as an enabling technology that serves as a basis for the development of application systems – such as a CAD system or a GIS. As noted by Güting [Güt94], "it is not claimed that a spatial DBMS is directly usable as an application-oriented GIS."

The data management requirements of spatial applications differ substantially from those of traditional business applications. Business applications tend to have simply structured data records, each occupying just a few bytes of memory. There is only a small number of relationships between data items, and transactions are comparatively short. Relational database systems meet these requirements extremely well. Their data model is table-oriented, therefore providing a natural fit to business requirements. By means of the transaction concept, one can check integrity constraints and reject inconsistencies.

With regard to spatial applications, however, conventional DBMS concepts are not nearly as efficient. Spatial databases contain multidimensional data with explicit knowledge about objects, their extent, and their position in space. The objects are usually represented in some vector-based format, and their relative position may be explicit or implicit (i.e., derivable from the internal representation of their absolute positions). To obtain a better understanding of the requirements in spatial database systems, let us first discuss some basic properties of spatial data:

1. Spatial data objects often have a *complex structure*. A spatial data object may be composed of a single point or several thousands of polygons, arbitrarily distributed across space. It is usually not possible to store collections of such objects in a single relational table with a fixed tuple size.

2. Spatial data is *dynamic*. Insertions and deletions are interleaved with updates, and data structures used in this context have to support this dynamic behavior without deteriorating over time.

3. Spatial databases tend to be *large*. Geographic maps, for example, typically occupy several gigabytes of storage. The seamless integration of secondary and tertiary memory is therefore essential for efficient processing [CDK+95].

4. There is *no standard algebra* defined on spatial data, although several proposals have been made in the past [SV90, Güt89, GS93]. This means in particular that there is no standardized set of base operators. The set of operators heavily depends on the given application domain.

5. Many spatial operators are *not closed*. The intersection of two polygons, for example, may return any number of single points, dangling edges, or disjoint polygons. This is particularly relevant when operators are applied consecutively.

6. Although computational costs vary, spatial operators are generally *more expensive* than standard relational operators.

7. Spatial data management suffers from an impedance mismatch between its theoretical requirements for *infinite accuracy* and the limited accuracy provided by computers.

Recent database research has helped to solve many related problems. This includes both extensions to the relational data model [SR86, KW87, HCL+90, SRH90, SPSW90, SSU91, Gaf96] and the development of flexible object-oriented approaches [Ore90b, WHM90, Deu90, SV92, BDK92, SCG+97]. We review some of the most relevant contributions in the remainder of this chapter. For more detailed information, the reader may consult the proceedings of the *Symposia on Spatial Databases (SSD)*, which have been held bi-annually since 1989 [BGSW90, GS91, AO93, EH95, SV97]. There are also several survey articles [GB90, Güt94, MP94] and textbooks [Sam90b, Sam90a, LT94, SVP+96] on the subject.

### 3.2.1 Spatial Data Types

An essential weakness in traditional commercial databases is that they do not provide any spatial data types. Following their orientation towards classical business applications, they may sometimes offer non-standard types such as *date* and *time* in addition to the classical data types *integer, real, character*, and *string*. Spatial data types, however, are not included in any of the standard commercial DBMS. On the other hand, such data types are a crucial requirement when it comes to processing geographic and environmental data.

For vector data, there have been several proposals on how to define a coherent and efficient spatial algebra [Güt89, SV92, Ege92, Ege94, GS95, Sch97]. It is generally assumed that the data objects are embedded in $d$-dimensional Euclidean space $E^d$ or a suitable subspace thereof. In this chapter, this space is also referred to as *universe* or *original space*. Any point object stored in a spatial database has a unique location in the universe, defined by its $d$ coordinates. Unless the distinction is essential, we use the term *point* both for locations in space and for point objects stored in the database. Note, however, that any point in space can be occupied by several point objects stored in the database.

A *(convex) d-dimensional polytope* $P$ in $E^d$ is defined as the intersection of some finite number of closed halfspaces in $E^d$, such that the dimension of the smallest affine subspace containing $P$ is $d$. If $a \in E^d - \{0\}$ and $c \in E^1$ then the $(d-1)$-dimensional set $H(a,c) = \{x \in E^d : x \cdot a = c\}$ defines a *hyperplane* in $E^d$. A hyperplane $H(a,c)$ defines two closed halfspaces, the *positive halfspace* $1 \cdot H(a,c) = \{x \in E^d : x \cdot a \geq c\}$, and the *negative halfspace* $-1 \cdot H(a,c) = \{x \in E^d : x \cdot a \leq c\}$. A hyperplane $H(a,c)$ *supports* a polytope $P$ if $H(a,c) \cap P \neq \emptyset$ and $P \subseteq 1 \cdot H(a,c)$, i.e., if $H(a,c)$ embeds parts of $P$'s boundary. If $H(a,c)$ is any hyperplane supporting $P$ then $P \cap H(a,c)$ is a *face* of $P$. The faces of dimension 1 are called *edges*; those of dimension 0 *vertices*.

By forming the union of some finite number of polytopes $Q_1, ..., Q_n$, one obtains a *(d-dimensional) polyhedron* $Q$ in $E^d$ that is not necessarily convex. Following the intuitive understanding of polyhedra, one usually requires that the $Q_i$ $(i = 1, ..., n)$ have to be connected. Note that this still allows for *polyhedra with holes*. Each face of $Q$ is either the face of some $Q_i$, or a fraction thereof, or the result of the intersection of two or more $Q_i$. Each polyhedron $P$ divides the points in space into three subsets that are mutually disjoint: its *interior*, its *boundary*, and its *exterior*.

One often uses the terms *line* and *polyline* to denote a one-dimensional polyhedron and the terms *polygon* and *region* to denote a two-dimensional polyhedron. If, for each $k$ $(0 \leq k \leq d)$, one views the set of $k$-dimensional polyhedra as a data type, one obtains the common collection of spatial data types $\{point, line, polygon, ...\}$. Combined types sometimes also occur. Curved objects can be obtained by extending the definitions given above.

An object in a spatial database is usually defined by several non-spatial attributes and one attribute of some spatial data type. This spatial attribute describes the object's *spatial extent*. Other common terms for spatial extent include *geometry, shape*, and *spatial extension*. For the *description* of the spatial extent, one finds the terms *shape descriptor/description, shape information,* and *geometric description,* among others.

As mentioned above, spatial data management suffers from an impedance mismatch between its theoretical requirements for infinite accuracy and the limited accuracy provided by computers. The fact that computers can not represent real numbers with arbitrary accuracy leads in fact to the implicit implementation of a grid-based data model. The representable numbers form the grid points, and all other points in space are rounded to a grid point nearby. If one does not take special precautions, this may lead to unpleasant consequences. For example, if a computer computes the intersection of two lines $l1$ and $l2$, it rounds the coordinates of the intersection point $(x, y)$ to some representable numbers nearby, say $(x', y')$. If one later checks whether the point $(x', y')$ lies on $l1$ or $l2$, the answer is likely to be negative.

Several authors have suggested grid-based data models to correct this fundamental flaw. Frank and collaborators have proposed an approach based on combinatorial topology [FK86, EFJ90]. Their technique revolves around the concept of the *simplicial complex,* which is a finite set of simplices such that the intersection of any two simplices in it is a face. As an alternative, Güting and Schneider have proposed the notion of a *realm* [GS93]. A realm is a finite set of points and lines over a discrete grid that fulfills certain conditions. In particular, each realm point has to be a grid point and each realm line has to be the direct connection between two realm points. Realm lines are not allowed to intersect. Figure 3.2 gives an example of a realm. The ROSE algebra proposed by the same authors is based on the same concept [GS95, Sch97].

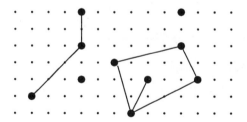

**Fig. 3.2.** Realm

## 3.2.2 Spatial Operators

As there exists neither a standard spatial algebra nor a standard spatial query language, there is also no consensus on a canonical set of spatial operators. Different applications use different operators, although some operators (such as intersection) are generally more common than others. Queries are often expressed by some extension of SQL that allows abstract data types to represent spatial objects and their associated operators; see Sect. 3.2.4 for a discussion. The result of a query is usually a set of spatial objects.

In the following, we give an informal definition of several common spatial database operators. The operators are grouped into six classes, depending on their respective input and output behavior (their *signature*). At least one of the operators has to be of a spatial data type. The input behavior refers to whether it is a unary, binary, or (in rare cases) $n$-ary $(n > 2)$ operator, as well as to the type of its operands. The output behavior refers to the type of result.

**Class U1: Unary operators with a boolean result.** These operators test a given spatial object for some property of interest. Examples include *triangle* or *convex*. They are often implemented ad hoc to serve some application-specific requirements. These operators are usually not part of the standard system architecture.

**Class U2: Unary operators with a scalar result.** These operators map a spatial object into a real or integer number. Examples include *dimension* or *volume*. Some operators in this class (such as *volume*) are an essential part of spatial information systems in general.

**Class U3: Unary operators with a spatial result.** The most important operators in this class are the *similarity operators* that map a spatial object into a similar object: *translation, rotation,* and *scaling.* Figure 3.3 shows a collection of similar triangles. Similarity operators are a subset of the broader class of *topological transformations,* where the dimension of the input object is always retained.

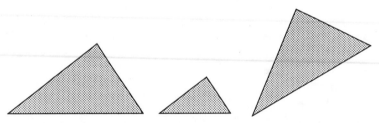

**Fig. 3.3.** Similar triangles

Yet other unary operators map a $d$-dimensional spatial object into an object of a lower or higher dimension. The *boundary* operator is an example of the former kind, mapping a $d$-dimensional spatial object $o$ $(d > 0)$ into a $(d-1)$-dimensional object that captures $o$'s boundary. Conversely, the *interior* operator maps a $d$-dimensional object into a $(d+1)$-dimensional object. For $d = 1$, for example, one obtains one or more polygons (Fig. 3.4) or, if the input line is not closed, an empty set.

To model the case where the result consists of several polygons, one has two options. Either one adopts a wider definition of the term *polyhedron* (and therefore also of *polygon*) that includes collections of disjoint components. Or one defines the operator to have a *set of polygons* as output. If the input line is self-intersecting, additional conventions are required to define the corresponding interior. A common convention is to include any region formed by the line, regardless of its orientation. Depending on the application domain, however, other conventions are common [GW89b].

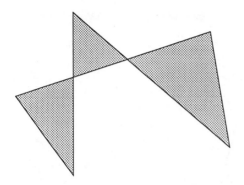

**Fig. 3.4.** Possible result of an *interior* operation

**Class B1: Binary operators with a boolean result.** These operators are known as *spatial predicates* or *spatial relationships.* They take two spatial objects (or sets of spatial objects) as input and produce a boolean as output. Several authors have tried to structure this range of operators [PE88, Ege90, Wor92, Güt94]. A common classification distinguishes between topological relationships, direction relationships, and metric relationships.

*Topological relationships* are invariant with respect to topological transformations, such as translation, rotation, and scaling. Examples are *intersects*, *contains*, *adjacent*, and *is_enclosed_by.*

*Direction relationships* refer to the current location of spatial objects and are therefore sensitive to some topological transformations, in particular rotations. Typical examples are *northwest_of* or *above.*

*Metric relationships* are sensitive to most topological transformations. They are in particular not invariant with respect to rotations and scalings. A typical example is *distance* < 10.

**Class B2: Binary operators with a scalar result.** These operators take two spatial objects (or sets of spatial objects) as input and compute an integer or real number as a result. The *distance* operator is a typical representative of this class.

**Class B3: Binary operators with a spatial result.** This set of operators comprises a wide range of different operators. Some of those most frequently used in practice include set operators and search operators.

The class of *set operators* comprises the union, difference, and intersection of two spatial objects or sets of objects. Notorious problems regarding these operators include the lack of a closure property (Fig. 3.5) and the handling of boundary phenomena (Fig. 3.6). The intersection in Fig. 3.5 is not a single polygon anymore. In Fig. 3.5 it is not clear a priori whether the user wants the line $AB$ to be part of the intersection. As can be seen from those examples, the result type depends not only on the type but on the shape and location of the input data.

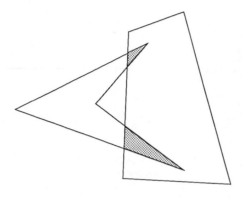

**Fig. 3.5.** Intersection of two polygons

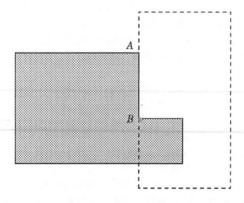

**Fig. 3.6.** Boundary problems after polygon intersection

    As with the *interior* operator, there are two ways to handle the closure problem. One option is to extend the definition of a *polyhedron* to include collections of disjoint components. In that case, the result of the intersection pictured in Fig. 3.5 would be considered as *one* object of type *polygon*. The other option is to define set operators to take two *sets* of spatial objects as input, and to produce one set of spatial objects as output. Of course, these sets may be empty or have one element only.

    The boundary problem is somewhat more difficult to solve because the desired result is highly application-dependent. There are many spatial database applications, such as computer-aided design, where the two operands have the same dimension $d$, and one usually wants the result to maintain this dimension. If one computes the intersection of two polygons, for example, one expects one (or several) *polygons* as a result. Dangling edges or vertices like those pictured in Fig. 3.6 are to be discarded. This strategy is supported by a slight variation of the standard set operators called *regularized* set operators [Til80]. These regularized operators first compute the standard union, intersection, or difference of the operands, then compute its interior, and finally add the boundary to the result. This way one always obtains a closed spatial object that has the same dimension as the two inputs. This approach, however, is not a panacea. There are numerous applications, especially in geographic and environmental information management, where the user is interested in all or at least some of the lower-dimensional parts of the result. It is therefore necessary to discuss those potential problems with the user before choosing a particular implementation.

Map overlays (cf. Fig. 3.1) are an important application of set operators. A map overlay is nothing but an intersection operation performed on two sets of polygons (the *maps*) [KBS91].

During the 1980s and 1990s, major progress has been made with regard to efficient algorithms for the computation of set operators. Especially the problem of computing polygon intersections has received a lot of attention in the computational geometry literature [PS85, Ede85, GW91].

The class of *search operators* concerns spatial search in a possibly large set of spatial objects. Formally speaking, a search operator takes two inputs. One is a set of spatial objects (the database), the other one is a single spatial object (the search pattern). Figure 3.7 illustrates some of those queries. The two most common search operations are the *point query* (Fig. 3.7a) and the *range query* (Fig. 3.7b). The point query asks for all spatial objects in the database that contain a given search point. The range query requests those objects that overlap a given search interval. Sometimes the search interval is replaced by a search object of arbitrary polygonal or polyhedral shape (*region query*, Fig. 3.7c). Common variations include the query for objects that are *near* a given search object, that are *adjacent* to a search object (*adjacency query*, Fig. 3.7d), that are *contained* in a given interval (*containment query*, Fig. 3.7e), or that *contain* a given interval (*enclosure query*, Fig. 3.7f). There has been a considerable amount of research on access methods to support the computation of search queries. Section 3.3 gives a detailed overview.

This concludes the classification of spatial operators. Table 3.1 gives a summary with selected examples. We use the term *d-Spatial* as a shorthand for a *d*-dimensional spatial object.

For the evaluation of spatial predicates (i.e., binary operators with a boolean result) in a database context, the *spatial join* operator has been introduced. In analogy to the classical join operation, it denotes the combination of two classes of spatial objects based on some spatial predicate. More formally, a spatial join takes two sets of spatial objects as input and produces

**a** point query

**b** range query

**c** region query

**d** adjacency query

**e** containment query

**f** enclosure query

**Fig. 3.7.** Spatial search queries [GG98]

**Table 3.1.** Examples of spatial operators

| Class | Operator | Operand | Result |
|-------|----------|---------|--------|
| U1 | *convex* | *d*-spatial | boolean |
| U2 | *dimension* | *d*-spatial | integer |
| U2 | *volume* | *d*-spatial | real |
| U3 | *boundary* | *d*-spatial | $(d-1)$-spatial |
| U3 | *interior* | *d*-spatial | $(d+1)$-spatial |

| Class | Operator | Operand-1 | Operand-2 | Result |
|-------|----------|-----------|-----------|--------|
| B1 | *intersects* | *d*-spatial | *d*-spatial | boolean |
| B1 | *neighbor* | *d*-spatial | *d*-spatial | boolean |
| B2 | *distance* | *d*-spatial | *d*-spatial | real |
| B3 | *regularized intersection* | *d*-spatial | *d*-spatial | *d*-spatial |
| B3 | *regularized union* | *d*-spatial | *d*-spatial | *d*-spatial |

a *set of pairs* of spatial objects as output, such that each pair fulfills the given spatial predicate. Examples include:

- Find all houses that are less than 10 km from a lake.
- Find all buildings that are located within a biotope.
- Find all schools that are more than 5 km away from a firestation.

All these queries combine one class of objects (e.g., *houses*) with another class of objects (e.g., *lakes*) and select those mixed pairs that fulfill the given predicate (e.g., *are less than 10 km from*). In a relational context, one can define the spatial join more formally as follows:

*The spatial join of two relations $R$ and $S$, denoted by $R \bowtie_{i\theta j} S$, is the set of tuples from $R \times S$ where the i-th column of $R$ and the j-th column of $S$ are of some spatial type, $\theta$ is some spatial predicate, and R.i stands in relation $\theta$ to S.j.*

This definition can be extended to object-oriented data models in a straightforward manner by viewing the tuples $r \in R$ and $s \in S$ as (spatial) objects and the relations $R$ and $S$ as sets of objects or class extents. As for the spatial predicate $\theta$, a brief survey of the literature on spatial joins [Ore86, Bec92, Rot91, Gün93, BKS93, BKSS94, LR94, AS94, GG95] yields a wide variety of possibilities, including *intersects*, *contains*, and *northwest*.

### 3.2.3 Implementation Issues

For the efficient computation of spatial operators one requires special implementations of the spatial data types mentioned above. Moreover, it is sometimes useful to represent the same spatial object in more than one way in order to represent a broad variety of spatial operators. A polygon, for example, may be represented as a vertex list – one of the most common representations. A vertex list is a list of the polygon's vertices, such as [(1,1),(5,1),(4,4)] for the triangle depicted in Fig. 3.3a. The vertex list is well suited to support similarity operators. A translation, for example, corresponds to the addition of the translation vector to each of the coordinates. A scaling (Fig. 3.3b) corresponds to a scalar multiplication, and a rotation (Fig. 3.3c) to a matrix multiplication. The vertex list representation of a polygon is also reasonably well suited to support set operators; it is the input representation of choice for most common algorithms.

Problems with that particular representation include the fact that it is *not unique*. The triangle described above could equally be described by the lists [(5,1),(4,4),(1,1)] and [(4,4),(1,1),(5,1)]. If one drops the implicit assumption that all listed points have to be vertices, one obtains an infinite number of alternative representations, such as [(2.5,2.5),(1,1),(5,1),(4,4)]. Furthermore, there are *no invariants* with respect to set operations; a translation, rotation, or scaling changes each single element of the representation. This also means that it is not easily possible to recognize whether two vertex lists represent congruent or similar polygons. Finally, it is not possible to recognize easily whether a vertex list corresponds to a self-intersecting line (Fig. 3.4) or whether it represents a polygon with holes. Depending on the current definition of a polygon, such vertex lists may be *invalid* representations.

Regardless of which representation is chosen, one has to map it into the data model of the given database. In classical relational environments, there are two options. First, the geometric representation may be hidden in a long field. This means that one of the columns in the relation is declared to have variable (in theory, infinite) length. The geometric representation is then stored in such a long field in a way that the application programs can interpret. The database system itself is usually not able to decode the representation. It is therefore not possible, for example, to ask SQL queries against that column. In the case of a vertex list, a typical relation *Polygon* could look as in Fig. 3.8.

Polygon

| ID | Color | Shape |
|------|-------|--------------------------|
| 2 | blue | [(1,1),(2,7),(3,9),(7,9)] |
| 4711 | red | [(2,1),(4,2),(1,5)] |
| ... | ... | ... |

**Fig. 3.8.** Relation with a long field

An alternative to this approach is to exclude the geometry completely from the database system and just give the name of the *external file* where it is stored; see Fig. 3.9 for an example.

Both approaches are somewhat problematic because they rely on software modules external to the database in order to interpret the representation. The database system itself does not have enough information to evaluate it, and it therefore has no concept of the geometry and topology of the stored objects. There is no database support for geometric operations. Furthermore, it

Polygon

| ID | Color | Shape |
|------|-------|-------------------|
| 2 | blue | /usr/john/pol2 |
| 4711 | red | /usr/john/pol4711 |
| ... | ... | ... |

**Fig. 3.9.** Relation with pointers to external files

is difficult to control redundancy because common components (e.g., shared corner points) cannot be extracted by the database. They are hidden somewhere within the long field. In the second option, the spatial information is not even protected by the standard database techniques of recovery and concurrency. It is not subject to transaction management because it is stored outside the database.

In a GIS context, the first option corresponds to the solution where the spatial data is stored inside a database but remains essentially a black box for the database management system. This approach is now favored by most major GIS, including ESRI's ARC/INFO and Siemens Nixdorf's SICAD. It also complies with the OpenGIS Simple Features Specifications proposed by the Open GIS Consortium (cf. Sect. 3.1.3). Many vendors provide efficient implementations of this design although the basic conceptual drawbacks remain.

*Abstract data types (ADTs)* provide a more robust, if not more efficient, way to integrate complex types into a database system. The basic idea is to encapsulate the implementation of a data type in such a way that one can communicate with instances of the data type only through a set of well-defined operators. The (interior) implementation of the data type and the associated operators are hidden to the (exterior) users. They have no way to review or modify those interior features.

The ADT concept can easily be adapted to the implementation of spatial data types and operators. It may also be used to give experienced users the opportunity to customize a DBMS according to their particular requirements, and to define those data types and operators that are most specific and appropriate for a given application. No DBMS could foresee all those needs. In order to appeal to a broad range of users, DBMS have to be designed independently of any particular application. With ADTs the system needs to offer only a small number of base data types and operators directly. A user may perceive only those base data types plus the ADTs that have been declared specifically for the given application. This view-like approach simplifies using the system, especially for inexperienced users, and helps to reduce training time.

In an SQL-style database environment, the embedding of the ADT concept is possible with minor syntax extensions. To define an ADT *Polygon*, for example, one has to provide some basic information about the implementation, such as the internal space requirements, the default value, and the routines used to input and output objects of that type.

The following example is based on a syntax proposed by Stonebraker and Rowe as part of their POSTGRES extensible database project [SR86]. We start with the definition of an abstract data type *Polygon*. The strings *char_to_polygon* and *polygon_to_char* refer to program modules that have to be called for various input and output tasks. The string *LongField* indicates that the representation may have variable length.

```
define type Polygon is
            (InternalLength = LongField,
             InputProc = char_to_polygon,
             OutputProc = polygon_to_char,
             Default = "")
```

Then one can define the desired operators, such as polygon intersection, or a spatial predicate that tests whether two given regions have the same area. Again, there are references to special program modules, such as *polygon_int* or *polygon_ae*. In addition to precedence and associativity information, the definition of the operator *area_equal* defines some special procedures for performing a selection or join operation that involves the operator.

```
define operator inter(Polygon,Polygon) returns Polygon is
            (Proc = polygon_int,
             Precedence = 3,
             Associativity = "left")

define operator area_equal(Polygon,Polygon) returns boolean is
            (Proc = polygon_ae,
             Precedence = 3,
             Associativity = "left",
             Hashes,
             Restrict = select_ae,
             Join = join_ae,
             Negator = not_polygon_ae)
```

After these preliminaries, one can use the ADT *Polygon* including its associated operators *inter* and *area_equal* just like any other data type. The following example shows how to define a relation *Map_Poly* for representing a thematic map. The operators perform an insertion (*append*), a modification (*update*), and a selection (*select*) that retrieves all polygons in *Map_Poly* whose area is identical to the area of the unit square.

```
create Map_Poly (ID = integer,
                 Layer = string,
                 Poly_Shape = Polygon)
```

```
append to Map_Poly (ID = 99,
                    Layer = "agriculture",
                    Poly_Shape = [(0,0),(2,3),(1,5)]

update Poly_Shape = inter(Poly_Shape,[(0,0),(4,1)])
from Map_Poly
where ID = 99

select ID
from Map_Poly
where area_equal(Poly_Shape,[(0,0),(1,0),(1,1),(0,1)])
```

Abstract data types have been shown to greatly enhance data security and to facilitate application programming. They can adapt to user requirements in a flexible manner by encapsulating data structures and operators of arbitrary complexity. Disadvantages of this approach include the duality of the connected programming paradigms: one always has to switch back and forth between the database-internal mode, which typically involves a non-procedural language such as SQL, and the external procedures, which are usually written in a procedural programming language. Furthermore, the internal structure of the data type is lost for the outside application; there is no way to retrieve any structural information from the ADT. This is a problem in particular for the database query optimizer. Without special accommodation, it is impossible for the optimizer to obtain any information about the complexity of the ADT operators that are included in a given query [GG95].

### 3.2.4 Spatial Query Languages

As we saw in the previous sections, any serious attempt to manage spatial data in a relational database framework requires some significant extensions at the logical and the physical level. These kinds of extension need to be supported at the query language level as well. Besides an ability to deal with spatial data types and operators, this involves in particular concepts to support the interactive working mode that is typical for many GIS/EIS applications. Pointing to objects or drawing on the screen with the mouse are typical examples of these dynamic interactions. Further extensions at the user interface level include [Voi95]: the graphical display of query results, including legends and labels; the display of unrequested context to improve readability; and the possibility of stepwise refinement of the display (logical zooming).

For many years, the database market has been dominated by a single query language: the Structured Query Language SQL. There has been a long discussion in the literature as to whether SQL is suitable for querying spatial databases. It was recognized early on that relational algebra and SQL alone are not able to provide this kind of support [Fra82, HR85, EF88, LM88].

In his 1992 paper "Why not SQL!" [Ege92], Egenhofer gives numerous examples of SQL's lack of expressive power and limitations of the relational model in general. With regard to the user interface level, Egenhofer notes the difficulties one encounters when trying to combine retrieval and display aspects in a single SQL query. Besides requiring specialized operators, this kind of combination usually leads to long and complex queries. The integration of selection by pointing (to the screen) is also problematic. There is no support in SQL for the stepwise refinement of queries, which is particularly important in a spatial database context. The underlying problem is that SQL does not provide a notion of state maintenance that allows users to interrupt their dialogue at a given point and resume their work later on.

Moreover, SQL does not support the notion of object identity in the presence of value changes. An object is defined only by its values. Object-oriented databases solve this problem by maintaining immutable object identifiers (cf. Sect. 3.4.1).

Finally, the relational model does not provide much support for meta-queries, i.e., queries referring to column names and other database schema information. Partly as a result of that, queries such as "What is this data item?," "Which unit of measurement is used for this item?," "What is the relation between the widths of I-95 and Highway 1?," or "What are possible soil classifications?" are difficult to frame in a relational context.

In some sense, however, with the stellar success of SQL the discussion about its appropriateness has become a moot point. The question is not whether SQL should be used – SQL is and will be used to query spatial databases as well. The question is rather which kind of extensions are desirable to optimize user-friendliness and performance of the resulting spatial data management system.

Various extensions to SQL have been proposed to deal specifically with spatial data, including PSQL [RL84, RL85], Spatial SQL [Ege91, Ege94], GEOQL [OMSD89, Ooi90], and the SQL-based GIS query languages for KGIS [IP87] and TIGRIS [HLS88]. Egenhofer [Ege92] gives a detailed overview of this work. Table 3.2 summarizes the features provided by those systems.

The key idea of Egenhofer's Spatial SQL is to integrate three different retrieval and display functionalities into a single framework but not into a single language:

1. Standard SQL is used for the querying and retrieval of nonspatial data, e.g.,

```
select Population
from   Town
where  Name = 'Orono'
```

**Table 3.2.** SQL extensions to handle spatial data [Ege94]

| Feature | GEO-QL | Ext'd SQL | PSQL | KGIS | TIG-RIS | Spatial SQL |
|---|---|---|---|---|---|---|
| Spatial ADT | + | +[1,2] | + | +[2] | +[2] | + |
| Graphical presentation | + | + | + | + | + | + |
| Result combination | - | - | - | +[3] | - | + |
| Context | - | - | - | +[3] | - | + |
| Content examination | - | - | - | - | - | + |
| Selection by pointing | + | - | - | + | - | + |
| Display manipulations | - | - | +[4] | - | - | + |
| Legend | - | - | - | - | - | + |
| Labels | - | - | + | - | - | + |
| Selection of map scale | - | - | + | - | - | + |
| Area of interest | - | - | - | + | - | + |

[1] Only spatial relationships
[2] No data definition
[3] Only for context
[4] As part of the picture list in the *on* clause

2. Certain extensions to SQL are used for the querying and retrieval of spatial data.
3. The graphical aspects are handled by a special Graphical Presentation Language GPL.

Spatial SQL preserves some but not all of SQL's key features. It maintains the *select-from-where* syntax. Unlike, for example, PSQL [RL84], it does not introduce any additional clauses such as *at* or *on*. As in SQL, the *where* clause in Spatial SQL refers to attribute values. For most queries, the result is a relation again. In order to avoid large and unwieldy queries, the display part of the language is strictly separated from the query and retrieval part. Spatial SQL does not provide any features for manipulating the display. All related problems are relayed to the presentation language GPL. This way, the number of actual extensions to standard SQL is kept low.

Spatial SQL supports selection by pointing. It introduces a special predicate *pick* that suspends the execution of the query and waits for interactive user input. For example, the query

```
select Name
from   County
where  Geometry = pick;
```

retrieves the name of the county the user points on. If one had written *City* instead of *County*, the result would have been the corresponding city, and so on.

To support the stepwise refinement of queries, Spatial SQL provides a rudimentary concept of state. Users can thus perform a sequence of queries, each of which refers to results obtained previously. Figure 3.10 shows a typical session. The user starts with the definition of the display environment using

GPL, then asks a Spatial SQL query that specifies the area and features of interest. This query may be refined in a stepwise manner by stating additional Spatial SQL queries that each refer to a result previously computed. Once the final result has been obtained, one may want to update the graphical presentation using GPL, ask another Spatial SQL query, and so on.

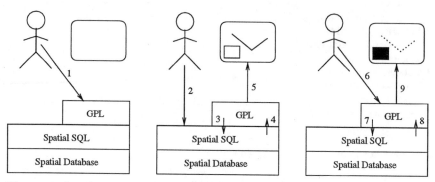

**Fig. 3.10.** Interaction with Spatial SQL and GPL [Ege94]

The graphical presentation language GPL offers a variety of choices for the graphic rendering of a query result. Its role can be compared to that of style files in LaTeX. Parameters include the scale of the rendering, the assignment of colors, patterns, and symbols to different thematic entities, the context to be displayed, and the display mode (e.g., alphanumeric, overlay, intersect, highlight). In order to avoid burdening the user with a large number of display details, the system manager can specify a number of standard settings as a starting point. Users can then modify those standard settings depending on their particular requirements.

To demonstrate the power and limitations of Spatial SQL, we present an example query similar to the one discussed in [Ege94]. Let us assume the user wants to see a specific thematic map based on a spatial database referring to Alameda County in Northern California. The spatial database contains information representing the cities, parcels, buildings, roads, rivers, and utility lines in that county. The user requests a map that helps him or her to decide on a planned real estate transaction. He or she asks for a map of Solano Street in Berkeley that includes all buildings (drawn using red dotted lines), parcel boundaries (green, dashed), and roads (blue, solid). The GPL commands to implement these directions are:

```
set legend
    color green
    pattern dashed
for select Boundary(Geometry)
    from   Parcel;
```

```
set legend
    color red
    pattern dotted
for select Geometry
    from    Building

set legend
    color blue
    pattern solid
for select Geometry
    from Road
```

The user can now identify the window of interest and set a context for the roads:

```
set window
    select Geometry
    from    Road
    where   Town.Name = ''Berkeley'';

set context
for Road.Geometry
    select Parcel.Geometry, Building.Geometry, Road.Name
    from    Road, Parcel, Building;
```

Having defined the thematic and graphical environment (in GPL), the user can now ask for a new map and enter the actual query (in Spatial SQL). Note the use of the spatial operator *inside*.

```
set mode new;
select    Road.Geometry
from      Road, Town
where     Town.Name = ''Berkeley''
and       Road.Name = ''Solano Street''
and       Road.Geometry inside Town.Geometry;
```

With further Spatial SQL commands the user may request more information about the objects displayed or manipulate the thematic or graphical environment. In order to point to a building and ask for its distance from the nearest firestation, for example, one may state the following query:

```
set mode alpha;
select    distance(Building.Geometry, Firestation.Geometry)
from      Building, Firestation Building
where     Building = pick
and       Firestation.Type = ''Fire Station'';
```

In summary, Spatial SQL represents an important step in the right direction. Several of Spatial SQL's features are suggestive of object-oriented concepts (see Sect. 3.4). Like many other SQL extensions, however, the approach seems somewhat ad hoc and lacks a formal foundation.

## 3.3 Multidimensional Access Methods

An important class of spatial operators that needs special support at the physical level is the class of *search operators* described above. Retrieval and update of spatial data is usually based not only on the value of certain alphanumeric attributes, but also on the spatial location of a data object. A retrieval query on a spatial database often requires the fast execution of a geometric search operation such as a point or range query. Both operations require fast access to those data objects in the database that occupy a given location in space.

To support such search operations, one needs special *multidimensional access methods*. This section gives an overview of the most important techniques in this area. It is based on a survey article by Gaede and Günther [GG98].

The main problem for the design of multidimensional access methods is that *there exists no total ordering among spatial objects that preserves spatial proximity.* In other words, there is no mapping from two- or higher-dimensional space into one-dimensional space such that any two objects that are spatially close in the higher-dimensional space are also close to each other in the one-dimensional sorted sequence.

This makes the design of efficient access methods in the spatial domain much more difficult than in traditional databases, where a broad range of efficient and well-understood access methods is available. Classical examples of such *one-dimensional access methods* include linear hashing [Lit80, Lar80], extendible hashing [FNPS79], and the B-tree [BM72, Com79]. These methods are an important foundation for almost all multidimensional access methods.

A natural approach to handle multidimensional search queries consists in the consecutive application of such single key structures, one per dimension. As Kriegel [Kri84] has pointed out, however, this approach can be quite inefficient. Since each index is traversed independently of the others, we cannot exploit the possibly high selectivity in one dimension for narrowing down the search in the remaining dimensions. Another interesting approach is to extend hashing simply by using a hash function that takes a $d$-dimensional vector as argument. A structure based on this idea is the grid file [NHS84]. Unfortunately this approach sometimes suffers from superlinear directory growth.

In Sect. 3.3.2 we provide a more detailed discussion of these and related problems. As these few examples already demonstrate, however, there is no easy and obvious way to extend single key structures in order to handle multidimensional data. On the other hand, there is a great variety of requirements that multidimensional access methods should meet, based on the properties of spatial data and their applications [Rob81, LS89a]:

1. *Dynamics.* As data objects are inserted and deleted from the database in any given order, access methods should continuously keep track of the changes.
2. *Secondary/tertiary storage management.* Despite growing main memories, it is often not possible to hold the complete database in main memory. Access methods therefore need to integrate secondary and tertiary storage in a seamless manner.
3. *Broad range of supported operations.* Access methods should not support just one particular type of operation (such as retrieval) at the expense of other tasks (such as deletion).
4. *Independence of the input data.* Access methods should maintain their efficiency even when the input data is highly skewed. This point is especially important for data that is distributed differently along the various dimensions.
5. *Simplicity.* Intricate access methods with special cases are often error-prone to implement and thus not sufficiently robust to be used in large-scale applications.
6. *Scalability.* Access methods should adapt well to growth in the underlying database.
7. *Time efficiency.* Spatial searches should be fast.
8. *Space efficiency.* An index should be small in size compared to the size of the data set.
9. *Concurrency and recovery.* In modern databases where multiple users concurrently update, retrieve, and insert data, access methods should provide robust techniques for transaction management without significant performance penalties.
10. *Minimum impact.* The integration of an access method into a database system should have minimum impact on existing parts of the system.

As for time efficiency, elapsed time is obviously what the user cares about, but one should keep in mind that the corresponding measurements depend greatly on implementation, hardware utilization, and other details. In the literature, one therefore often finds a seemingly more objective performance measure: the number of disk accesses performed during a search. This approach, which has become popular with the B-tree, is based on the assumption that most searches are I/O-bound rather than CPU-bound – an assumption that is not always true in spatial data management, however. In applications where objects have complex shapes, the refinement step can incur major CPU costs and change the balance with I/O [Gae95b, HS95]. Of course, one should keep the minimization of disk accesses in mind as *one* design goal. Practical evaluations, however, should always give some information on elapsed times and the conditions under which they were achieved. A major design goal of multidimensional access methods is to meet the performance characteristics of one-dimensional B-trees: access methods should guarantee a logarithmic worst-case search performance for *all* possible input data distributions regardless of the insertion sequence.

A common approach to meet the requirements listed above consists in a two-step approximation approach (Fig. 3.11). The idea is to abstract from the actual shape of a spatial object before inserting it into an index. This can be achieved by approximating the original data object with a simpler shape, such as a bounding box or a sphere. Given the minimum bounding interval $I_i(o) = [l_i, u_i]$ $(l_i, u_i \in E^1)$ describing the extent of the spatial object $o$ along dimension $i$, the $d$-dimensional *minimum bounding box (MBB)* is defined by $I^d(o) = I_1(o) \times I_2(o) \times ... \times I_d(o)$.

An index may only administer the MBB of each object, together with a pointer to the description of the object's database entry (the *object ID* or *object reference*). With this design, the index only produces a set of *candidate solutions*. This step is therefore termed the *filter step*. For each element of that candidate set we have to decide whether the MBB is sufficient to decide that the actual object *must* indeed satisfy the search predicate. In those cases, the object can be added directly to the query result. However, there are often cases where the MBB does not prove to be sufficient. In a *refinement step* we then have to retrieve the exact shape information from secondary memory and test it against the predicate. If the predicate evaluates to true, the object is added to the query result as well, otherwise we have a *false drop*.

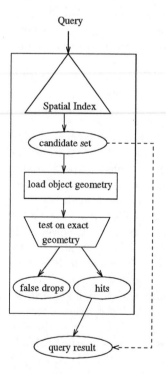

Fig. 3.11. Multi-step spatial query processing [BKSS94]

We have already introduced the term *multidimensional access methods* to denote the large class of access methods that support searches in spatial databases and that are the subject of this chapter. Within this class, we distinguish between *point access methods (PAMs)* and *spatial access methods (SAMs)*. Point access methods have primarily been designed to perform spatial searches on point databases. The points may be embedded in two or more dimensions, but they do not have a spatial extension. Spatial access methods, on the other hand, are able to manage extended objects, such as lines, polygons, or even higher-dimensional polyhedra. In the literature, one often finds the term *spatial access method* referring to what we call *multidimensional access method*. Other terms used for this purpose include *spatial index* or *spatial index structure*.

## 3.3.1 Data Structures for Main Memory

Early multidimensional access methods did not account for paged secondary memory and are therefore less suited for large spatial databases. In this section, we review several of these fundamental data structures, which are adapted and incorporated in numerous multidimensional access methods. To illustrate the methods, we introduce a small scenario that we shall use as a running example throughout this chapter. The scenario, depicted in Fig. 3.12, contains ten points $p_i$ $(i = 1, ..., 10)$ and ten polygons $r_i$, randomly distributed in a finite two-dimensional universe. To represent polygons, we shall often use their centroids $c_i$ (not pictured) or their minimum bounding boxes (MBBs) $m_i$. Note that the quality of the MBB approximation varies considerably. The MBB $m_8$, for example, provides a fairly tight fit, whereas $r_5$ is only about half as large as its MBB $m_5$.

*The k-d-Tree [Ben75, Ben79].* The k-d-tree is a $d$-dimensional binary search tree that represents a recursive subdivision of the universe into subspaces by means of $(d-1)$-dimensional hyperplanes. The hyperplanes are iso-oriented, and their direction alternates between the $d$ possibilities. For $d = 3$, for example, splitting hyperplanes are alternately perpendicular to the $x$-axis, the $y$-axis, and the $z$-axis. Each splitting hyperplane has to contain at least one data point, which is used for its representation in the tree. Interior nodes have one or two descendants each and function as discriminators to guide the search. Searching and insertion of new points are straightforward operations. Deletion is somewhat more complicated and may cause a reorganization of the subtree below the data point to be deleted.

Figure 3.13 shows a k-d-tree for the running example. Because the tree can only handle points, we represent the polygons by their centroids $c_i$. The first splitting line is the vertical line crossing $c_3$. We therefore store $c_3$ in the root of the corresponding k-d-tree. The next splits occur along horizontal lines crossing $p_{10}$ (for the left subtree) and $c_7$ (for the right subtree), and so on.

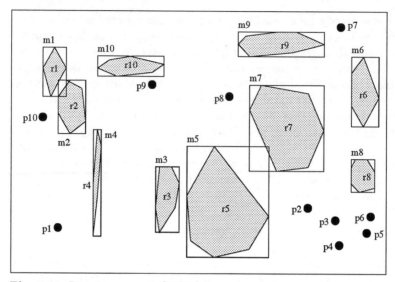

**Fig. 3.12.** Running example [GG98]

One disadvantage of the k-d-tree is that the structure depends on the order in which the points are inserted. Another one is that data points are scattered all over the tree. The *adaptive k-d-tree* [BF79] mitigates these problems by choosing a split such that one finds about the same number of elements on both sides. While the splitting hyperplanes are still parallel to the axes, they do not have to contain a data point and their directions no longer have to alternate strictly. As a result, the split points are not part of the input data; all data points are stored in the leaves.

A disadvantage common to all k-d-trees is that for certain distributions no hyperplane can be found that splits the data points in an even manner [LS89a]. By introducing a more flexible partitioning scheme, the BSP tree presented next avoids this problem completely.

*The BSP Tree [FKN80, FAG83].* Splitting the universe only along iso-oriented hyperplanes is a severe restriction. Allowing arbitrary orientations gives more flexibility to find a hyperplane that is well suited for the split. A well-known example of such a method is the BSP (Binary Space Partitioning) tree. Like k-d-trees, BSP trees are binary trees that represent a recursive subdivision of the universe into subspaces by means of $(d-1)$-dimensional hyperplanes. Each subspace is subdivided independently of its history and of the other subspaces. The choice of the partitioning hyperplanes depends on the distribution of the data objects in a given subspace. The decomposition usually continues until the number of objects in each subspace is below a given threshold.

The resulting partition of the universe can be represented by a BSP tree, where each hyperplane corresponds to an interior node of the tree and each

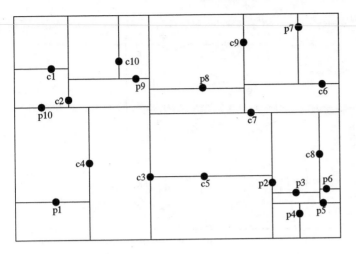

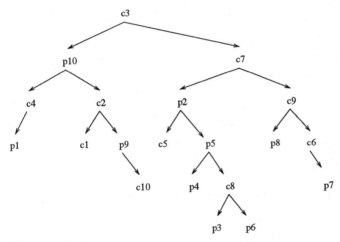

**Fig. 3.13.** k-d-tree

subspace corresponds to a leaf. Each leaf stores references to objects contained in the corresponding subspace. Figure 3.14 shows a BSP tree for the running example with no more than two objects per subspace.

In order to perform a point search, we insert the search point into the root of the tree and determine on which side of the corresponding hyperplane the point is located. Next, we insert the point into the corresponding subtree and proceed recursively until we reach a leaf of the tree. Finally, we have to examine the objects in the corresponding subspace to see whether they contain the search point. The range search algorithm is a straightforward generalization.

BSP trees can adapt well to different data distributions. However, they are typically not balanced and may have very deep subtrees, which has a negative

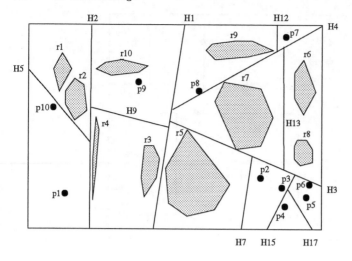

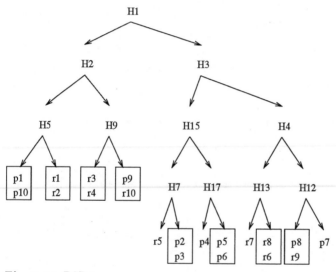

**Fig. 3.14.** BSP tree

impact on tree performance. BSP trees also have higher space requirements, since storing an arbitrary hyperplane per split occupies more storage space than a discriminator, which is typically just a real number.

*The Quadtree.* The quadtree with its many variants is a close relative of the k-d-tree. For an extensive discussion of this structure, see [Sam84, Sam90b, Sam90a]. While the term quadtree usually refers to the two-dimensional variant, the basic idea applies to arbitrary dimensions. Like the k-d-tree, the quadtree decomposes the universe by means of iso-oriented hyperplanes. An important difference, however, is the fact that quadtrees are not binary trees anymore. In $d$ dimensions, the interior nodes of a quadtree have $2^d$ descendants, each corresponding to an interval-shaped partition of the given subspace. These partitions do not have to be of equal size, although that is often the case. For $d = 2$, for example, each interior node has four descendants, each corresponding to a rectangle. These rectangles are typically referred to as the NW, NE, SW, and SE (northwest etc.) quadrants. The decomposition into subspaces usually continues until the number of objects in each partition is below a given threshold. Quadtrees are therefore not necessarily balanced; subtrees corresponding to densely populated regions may be deeper than others.

Searching in a quadtree is similar to searching in an ordinary binary search tree. At each level, one has to decide which of the four subtrees need to be included in the future search. In the case of a point query, typically only one subtree qualifies, whereas for range queries there are often several. We repeat this search step recursively until we reach the leaves of the tree.

Finkel and Bentley [FB74] proposed one of the first variants. It is called a *point quadtree* and is essentially a multidimensional binary search tree. The point quadtree is constructed consecutively by inserting the data points one by one. For each point, we first perform a point search. If we do not find the point in the tree, we insert it into the leaf node where the search has terminated. The corresponding partition is divided into $2^d$ subspaces with the new point at the center. The deletion of a point requires the restructuring of the subtree below the corresponding quadtree node. A simple way to achieve this is to reinsert all points in the subtree. Figure 3.15 shows a two-dimensional point quadtree for the running example.

Another popular variant is the *region quadtree* [Sam84]. Region quadtrees are based on a *regular decomposition* of the universe, i.e., the $2^d$ subspaces resulting from a partition are always of equal size. This greatly facilitates searches. For the running example, Fig. 3.16 shows how region quadtrees can be used to represent sets of points. Here the threshold for the number of points in any given subspace was set to one. There are many other variants of the quadtree concept; the books by Samet [Sam90a, Sam90b] give a comprehensive survey.

## 3.3.2 Point Access Methods

The multidimensional data structures presented in the previous section do not take secondary storage management explicitly into account. They were originally designed for main memory applications where all the data is available without accessing the disk. Despite growing main memories, this is of course not always the case. In many spatial database applications the amount of data to be managed is notoriously large. While one can certainly use main memory structures for data that resides on disk, their performance will often be considerably below the optimum because there is no control over the way

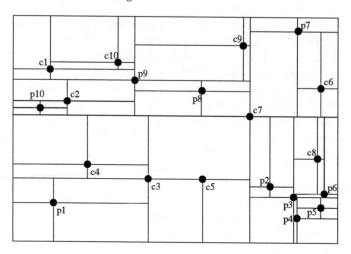

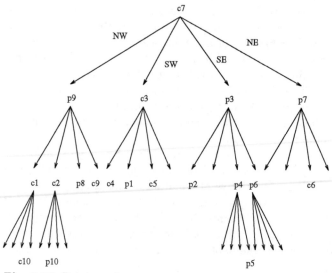

**Fig. 3.15.** Point quadtree

the operating system performs the disk accesses. The access methods presented in this and the following section have been designed with secondary storage management in mind. Their operations are closely coordinated with the operating system to ensure that overall performance is optimized.

As mentioned before, we shall first present a selection of *point access methods (PAMs)*. Usually, the points in the database are organized in a number of buckets, each of which corresponds to a disk page and to some subspace of the universe. The subspaces (often referred to as *data regions, bucket regions,* or simply *regions*, even though their dimension may be greater than two)

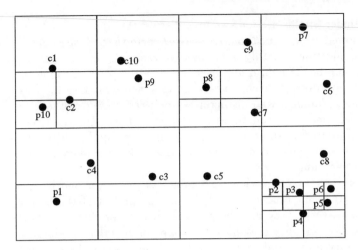

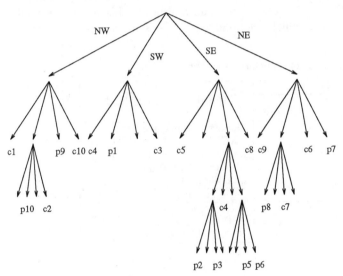

**Fig. 3.16.** Region quadtree

need not be rectilinear, although they often are. The buckets are accessed by means of a search tree or some $d$-dimensional hash function.

The grid file [NHS84], for example, uses a directory and a grid-like partition of the universe to answer an exact match query with exactly two disk accesses. Furthermore, there are multidimensional hashing schemes [Tam82, KS86, KS88], multilevel grid files [WK85, HSW88b], and hash trees [Ouk85, Oto85], which organize the directory as a tree structure. Tree-based access methods are usually a generalization of the B-tree to higher dimensions, such as the k-d-B-tree [Rob81] or the hB-tree [LS89a].

In the remainder of this section, we first discuss the approaches based on *hashing*, followed by *hierarchical (tree-based) methods* and *space-filling curves*. This classification is hardly unambiguous, especially in the presence of an increasing number of hybrid approaches that attempt to combine the advantages of several different techniques. Our approach resembles the classification of Samet [Sam90a] who distinguishes between *bucket* methods on the one hand and *hierarchical* methods on the other hand. His discussion of the latter is primarily in the context of main memory applications. Our presentation focuses throughout on structures that take secondary storage management into account.

Another interesting taxonomy has been proposed by Seeger and Kriegel [SK90] who classify point access methods by the properties of the bucket regions (Table 3.3). First, they may be pairwise *disjoint* or they may have mutual overlaps. Second, they may have the shape of an *interval* (box) or be of some arbitrary polyhedral shape. Third, they may cover the *complete* universe or just those parts that contain data objects. This taxonomy results in eight classes, four of which are populated by existing access methods.

**Table 3.3.** Classification of point access methods [SK90]

| Property | | | Point access method |
|-----------|----------|----------|---------------------|
| Intervals | Complete | Disjoint | |
| × | × | × | Quadtree, k-d-B-tree, EXCELL, grid file, MOLHPE, interpolation-based grid file, LSD tree, PLOP hashing, quantile hashing |
| × | × | | Twin grid file |
| × | | × | Buddy tree |
| | × | × | BSP tree, BD-tree, BANG file, hB-tree |

**Multidimensional Hashing.** Although there is no total order for objects in two- and higher-dimensional space that completely preserves spatial proximity, there have been numerous attempts to construct hashing functions that preserve proximity at least to some extent. The goal of all these heuristics is that objects located close to each other in original space should be stored close together on the disk with high probability. This could contribute substantially to minimizing the number of seek operations per range query.

*The Grid File [NHS84].* The grid file superimposes a $d$-dimensional orthogonal *grid* on the universe. Because the grid is not necessarily regular, the resulting *cells* may be of different shapes and sizes. A *grid directory* associates one or more of these cells with *data buckets*, which are stored on one disk page each. Each cell is associated with one bucket, but a bucket may contain several adjacent cells. Since the directory may grow large, it is usually kept on secondary storage. To guarantee that data items are always found with no more than two disk accesses for exact match queries, the grid itself is kept in main memory, represented by $d$ one-dimensional arrays called *scales*.

Figure 3.17 shows a grid file for the running example. We assume bucket capacity to be four data points. The center of the figure shows the root directory with scales on the $x$- and $y$-axes. The data points are displayed in the directory for demonstration purposes only; they are not stored there, of course. In the lower left part, four cells are combined into a single bucket, represented by four pointers to a single page. Hence there are four directory entries for the same page, which illustrates a well-known problem of the grid file: the growth of the directory may be superlinear, even for data that is uniformly distributed [Reg85, Wid91]. The bucket region containing the point $c_5$ could have been merged with one of the neighboring buckets for better storage utilization. Combining such *buddies* may cause problems later, however, when trying to combine buckets that are underoccupied [Hin85, SK90]. This is a trade-off that has to be solved depending on the particular application at hand.

To answer an exact match query, one first uses the scales to locate the cell containing the search point. If the appropriate grid cell is not in main memory, one disk access is necessary. The loaded cell contains a reference to the page where possibly matching data may be found. Retrieving this page may require another disk access. Altogether, no more than two page accesses are necessary to answer the query. For a range query, one has to examine all cells that overlap the search interval. After eliminating duplicates, one fetches the corresponding data pages into memory for a more detailed inspection.

To insert a point, one first performs an exact match query to locate the cell and the data page $\nu_i$ where the entry should be inserted. If there is sufficient space left on $\nu_i$, the new entry is inserted. If not, a splitting hyperplane $H$ is introduced and a new data page $\nu_j$ is allocated. The new entry and the entries of the original page $\nu_i$ are redistributed among $\nu_i$ and $\nu_j$, depending on their location relative to $H$. $H$ is inserted into the corresponding scale; all cells that intersect $H$ are split accordingly. Splitting is therefore not a local operation and can lead to superlinear directory growth [Reg85, Fre87, Wid91].

Deletion is not a local operation either. With the deletion of an entry, the storage utilization of the corresponding data page may drop below the given threshold. Depending on the current partitioning of space, it may then be possible to merge this page with a neighboring page and to drop the partitioning hyperplane from the corresponding scale. Depending on the implementation of the grid directory, merging may require a complete directory scan [Hin85]. Hinrichs [Hin85] discusses several methods for finding candidates with which a given data bucket can merge, including the *neighbor system* and the *multidimensional buddy system*. The neighbor system allows the merging of two adjacent regions as long as the result is a rectangular region again. In the

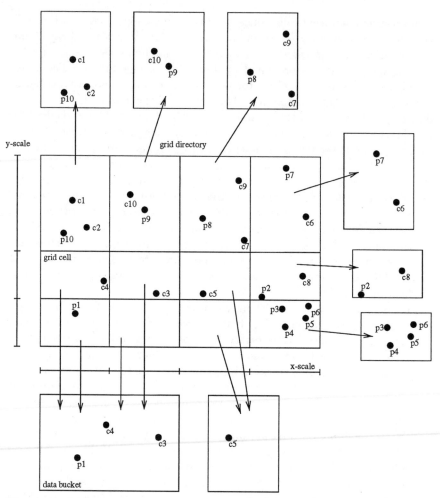

**Fig. 3.17.** Grid file

buddy system, two adjacent regions can be merged provided that the joined region can be obtained by a regular binary subdivision of the universe. Neither system is able to eliminate completely the possibility of a *deadlock*, where no merging is feasible because the resulting bucket region would not be box-shaped [Hin85, SK90].

*Multidimensional Linear Hashing.* Unlike multidimensional extendible hashing, multidimensional *linear* hashing uses no or only a very small directory. It therefore occupies relatively little storage compared to extendible hashing, and it is usually possible to keep all relevant information in main memory.

Several different strategies have been proposed to perform the required address computation. While early proposals [OS83b] failed to support range queries, Kriegel and Seeger [KS86] later proposed a variant of linear hashing called *multidimensional order-preserving linear hashing with partial expansions (MOLHPE)*. This structure is based on the idea of partially extending the buckets without expanding the file size at the same time. To this end, they use a *d*-dimensional expansion pointer referring to the group of pages that will be expanded next. With this strategy, Kriegel and Seeger can guarantee a modest file growth, at least in the case of well-behaved data. According to their experimental results, MOLHPE outperforms its competitors for uniformly distributed data. It fails, however, for non-uniform distributions, mostly because the hashing function does not adapt gracefully to the given distribution.

To solve this problem, the same authors later applied a stochastic technique called $\alpha$-quantiles [Bur84] to determine the split points. The resulting access method was called *quantile hashing* [KS87, KS89]. The critical property of the division in quantile hashing is that the original data, which may have a non-uniform distribution, is transformed into uniformly distributed values for $\alpha$. These values are then used as input to the MOLHPE algorithms for retrieval and update. Since the region boundaries are not necessarily simple binary intervals, a small directory is needed. In exchange, skewed input data can be maintained as efficiently as uniformly distributed data. A further variant called *PLOP (piecewise linear order-preserving) hashing* was proposed by the same authors one year later [KS88].

A related method with better order-preserving properties than MOLHPE has been reported by Hutflesz et al. [HSW88a]. Their *dynamic z-hashing* uses a space-filling technique called *z-ordering* [OM84] to guarantee that points located close to each other are also stored close together on the disk. Z-ordering is described in detail in Sects. 3.3.2 and 3.3.3. One disadvantage of z-hashing is that a number of useless data blocks will be generated, as in the interpolation-based grid file [Ouk85]. On the other hand, z-hashing allows reading of three to four buckets in a row, on average, before a seek is required, whereas MOLHPE manages to read only one [HSW88a]. Widmayer [Wid91] later noted, however, that both z-hashing and MOLHPE are of limited use in practice due to their inability to adapt to different data distributions.

**Hierarchical Point Access Methods.** After discussing PAMs based on hashing we now turn to PAMs based on a binary or multiway tree structure. Except for the BANG file and the buddy tree, which are hybrid structures, they do not perform any address computation. Like hashing-based methods, however, they organize the data points in a number of buckets. Each bucket usually corresponds to a leaf node of the tree (also called *data node*) and a disk page, which contains those points that are located in the corresponding bucket region. The *interior nodes* of the tree (also called *index nodes*) are used to guide the search; each of them typically corresponds to a larger subspace of the universe that contains all bucket regions in the subtree below. A search operation is then performed by a top-down tree traversal.

Differences between tree structures are mainly based on the characteristics of the regions. Table 3.3 shows that in most PAMs the regions at the same tree level form a partitioning of the universe, i.e., they are mutually disjoint, with their union being the complete space. For SAMs this is not necessarily true. As we will see in Sect. 3.3.3, overlapping regions and partial coverage are important techniques for improving the search performance of SAMs.

*The k-d-B-Tree [Rob81].* The k-d-B-tree combines the properties of the adaptive k-d-tree and the B-tree for handling multidimensional points. It partitions the universe in the manner of an adaptive k-d-tree and associates the resulting subspaces with tree nodes. Each interior node corresponds to an interval-shaped region. Regions corresponding to nodes at the same tree level are mutually disjoint; their union is the complete universe. The leaf nodes store the data points that are located in the corresponding partition. Like the B-tree, the k-d-B-tree is a perfectly balanced tree that adapts well to the distribution of the data. Unlike B-trees, however, no minimum space utilization can be guaranteed. A k-d-B-tree for the running example is sketched in Fig. 3.18.

Search queries are answered in a straightforward manner, analogously to the k-d-tree algorithms. For the insertion of a new data point, one first performs a point search to locate the right bucket. If it is not full, the entry is inserted. Otherwise, it is split and about half of the entries are shifted to the new data node. In order to find an optimal split, various heuristics are available [Rob81]. If the parent index node does not have enough space left to accommodate the new entries, a new page is allocated and the index node is split by a hyperplane. The entries are distributed among the two pages, depending on their position relative to the splitting hyperplane, and the split is propagated up the tree. The split of the index node may also affect regions at *lower* levels of the tree, which have to be split by this hyperplane as well. Because of this *forced split* effect, it is not possible to guarantee a minimum storage utilization.

Deletion is straightforward. After performing an exact match query, the entry is removed. If the number of entries drops below a given threshold, the data node may be merged with a sibling data node as long as the union remains a $d$-dimensional interval. The procedure for finding a suitable sibling node to merge with may involve several nodes. The union of data pages results in the deletion of at least one hyperplane in the parent index node. If an underflow occurs, the deletion has to be propagated up the tree.

*The Buddy Tree [SK90].* The buddy tree is a dynamic hashing scheme with a tree-structured directory. The tree is constructed by consecutive insertion, cutting the universe recursively into two parts of equal size with iso-oriented hyperplanes. Each interior node $\nu$ corresponds to a $d$-dimensional partition $P^d(\nu)$ and to an interval $I^d(\nu) \subseteq P^d(\nu)$. $I^d(\nu)$ is the MBB of the points or intervals below $\nu$. Partitions $P^d$ (and therefore intervals $I^d$) that correspond to nodes on the same tree level are mutually disjoint. As in all tree-based struc-

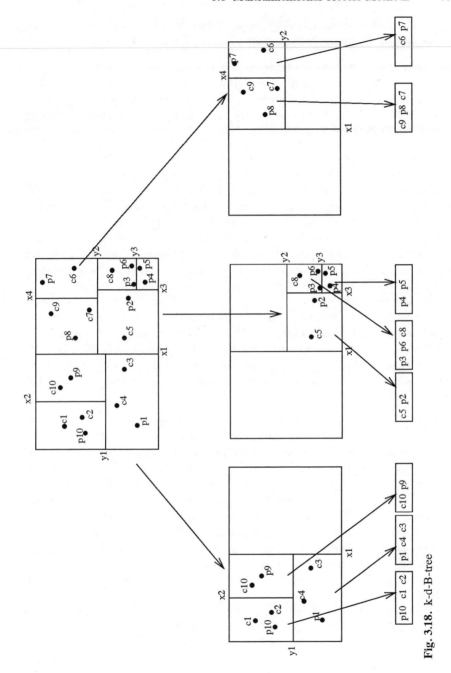

**Fig. 3.18.** k-d-B-tree

tures, the leaves of the directory point to the data pages. Other important properties of the buddy tree include:

1. Each directory node contains at least two entries.
2. Whenever a node $\nu$ is split, the MBBs $I^d(\nu_i)$ and $I^d(\nu_j)$ of the two resulting subnodes $\nu_i$ and $\nu_j$ are recomputed to reflect the current situation.
3. Except for the root of the directory, there is exactly one pointer referring to each directory page.

Due to property 1, the buddy tree may not be balanced, i.e., the leaves of the directory may be on different levels. Property 2 is intended to achieve a high selectivity at the directory level. Properties 1 and 3 ensure that the growth of the directory remains linear. To avoid the deadlock problem of the grid file, the buddy tree uses k-d-*tries* [Ore82] to partition the universe. Only a restricted number of buddies are admitted, namely those that could have been obtained by some recursive halving of the universe. However, as shown by Seeger and Kriegel [SK90], the number of possible buddies is larger than in the grid file and other structures, which makes the buddy tree more flexible in the case of updates.

Experiments by Kriegel et al. [KSSS90] indicate that the buddy tree is superior to several other PAMs, including the hB-tree, the BANG file, and the two-level grid file. A buddy tree for the running example is shown in Fig. 3.19. Seeger [See91] later showed that the buddy tree can be easily modified to handle spatially extended objects by using one of the techniques presented in Sect. 3.3.3.

*The BANG File [Fre87].* To obtain a better adaption to the given data points, Freeston proposed a structure called the BANG (Balanced And Nested Grid) file – even though it differs from the grid file in many aspects. Like the grid file, it partitions the universe into intervals (boxes). What is different, however, is that in the BANG file, bucket regions may intersect, which is not possible in the regular grid file. In particular, one can form non-rectangular bucket regions by taking the geometric difference of two or more intervals *(nesting)*. To increase storage utilization, it is possible during insertion to redistribute points between different buckets. To manage the directory, the BANG file uses a balanced search tree structure. In combination with the hash-based partitioning of the universe, the BANG file can therefore be viewed as a hybrid structure.

Figure 3.20 shows a BANG file for the running example. Three rectangles have been cut out of the universe $R_1$: $R_2$, $R_5$, and $R_6$. In turn, the rectangles $R_3$ and $R_4$ are nested into $R_2$ and $R_5$, respectively.

In order to achieve a high storage utilization, the BANG file performs spanning splits, which may lead to the displacement of parts of the tree. As a result, a point search may require in the worst case a traversal of the entire directory in a depth-first manner. To address this problem, Freeston [Fre90a]

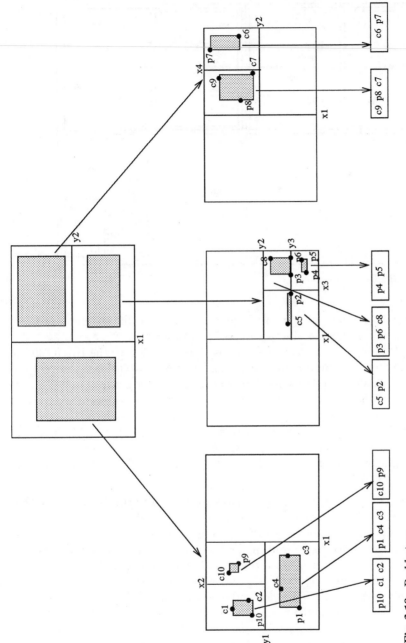

**Fig. 3.19.** Buddy tree

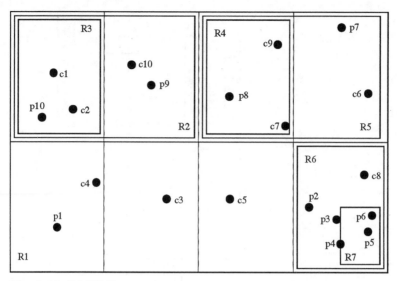

**Fig. 3.20.** BANG file

later proposed different splitting strategies that avoid the spanning problem at the possible expense of lower storage utilization.

In [Fre90b], the same author proposes an extension to the BANG file to handle extended objects. As often found in PAM extensions, the centroid is used to determine the bucket in which to place a given object. To take account of the object's spatial extension, the bucket regions are extended where necessary [Ooi90, SK88].

The BD-tree [OS83a] is an early precursor of the BANG file. It stores binary encodings (called *DZ-expressions*) of excised data regions in a binary tree. Unfortunately, the authors do not describe how the directory is maintained on disk.

*The hB-Tree [LS89a, LS90].* The hB-tree is related to the k-d-B-tree in that it utilizes k-d-trees to organize the space represented by its interior nodes. One of the most noteworthy differences is that node splitting is based on multiple attributes. As a result, nodes no longer correspond to $d$-dimensional intervals, but to intervals from which smaller intervals have been excised. As with the BANG file, the result is a somewhat fractal structure (a *holey brick*) with an external *enclosing region* and several cavities called *extracted regions*. This technique avoids the cascading of splits that is typical for many other structures.

To minimize redundancy, the k-d-tree corresponding to an interior node can have several leaves pointing to the same child node. Strictly speaking, the hB-tree is therefore no longer a tree but a directed acyclic graph. With regard to the geometry, this corresponds to the union of the corresponding regions. Once again, the resulting region is typically no longer box-shaped.

This peculiarity is illustrated in Fig. 3.21, which shows an hB-tree for the running example. Here the root node contains two pointers to its left descendant node. Its corresponding region $u$ is the union of two rectangles: the one to the left of $x_1$ and the one above $y_1$. The remaining space (the right lower quadrant) is excluded from $u$, which is made explicit by the entry *ext* in the corresponding k-d-tree. A similar observation applies to region $G$, which is again L-shaped: it corresponds to the NW, the SE, and the NE quadrant of the rectangle above $y_1$.

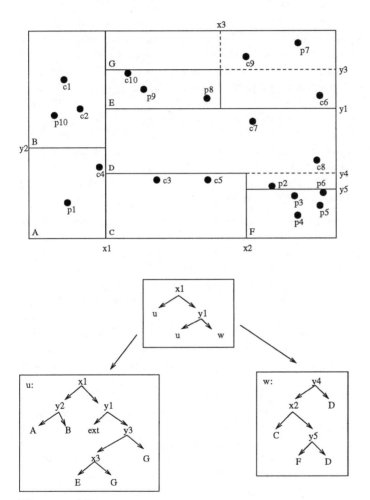

**Fig. 3.21.** hB-tree

Searching is similar to the k-d-B-tree; each internal k-d-tree is traversed as usual. The search performance of the one-dimensional hB-tree is competitive with the B-tree. Insertions are also carried out analogously to the k-d-B-tree until a leaf node reaches its capacity and a split is required. Instead of using just a single hyperplane to split the node, the hB-tree split is based on more than one attribute and on the internal k-d-tree of the data node to be split. Lomet and Salzberg [LS89a] show that this policy guarantees a worst-case data distribution between the two resulting two nodes of $1/3 : 2/3$. This observation is not restricted to the hB-tree but generalizes to other access methods such as the BD-tree and the BANG file.

The split of the leaf node causes the introduction of an additional k-d-tree node to describe the resulting subspace. This may in turn lead to the split of the ancestor node and its k-d-tree. Since k-d-trees are not height-balanced, splitting the tree at its root may lead to an unbalanced distribution of the nodes. The tree is therefore usually split at a lower level, which corresponds to the excision of a convex region from the space corresponding to the node to be split. The entries belonging to that subspace are extracted and moved to a new hB-tree node. To reflect the absence of the excised region, the hB-tree node is assigned a special marker, which indicates that the region is no longer a simple interval (cf. the *ext* node in Fig. 3.21). With this technique the problem of forced splits is avoided, i.e., splits are local and do not have to be propagated downwards.

A later paper by Evangelidis et al. [ELS95] discusses extensions to the hB-tree to allow for concurrency and recovery.

**Space-Filling Curves.** We already mentioned the main reason why the design of multidimensional access methods is so difficult compared to the one-dimensional case: there is no total order that preserves spatial proximity. One way out of this dilemma is to find heuristic solutions, i.e., to look for total orders that preserve spatial proximity at least to some extent. The idea is that if two objects are located close together in the original space, there should *at least be a high probability* that they are close together in the total order, i.e., in the one-dimensional image space. For the organization of this total order one could then use a one-dimensional access method (such as a $B^+$-tree), which may provide good performance at least for some spatial queries.

Research on the underlying mapping problem goes back well into the last century; see [Sag94] for a survey. With regard to its relevance for spatial searching, Samet [Sam90b] provides a good overview of the subject. One thing all proposals have in common is that they first partition the universe with a grid. Each of the grid cells is labeled with a unique number that defines its position in the total order (the *space-filling curve*). The points in the given data set are then sorted and indexed according to the grid cell they are contained in. Note that while the labeling is independent of the given data, it is obviously critical for the preservation of proximity in one-dimensional address space. That is, the way we label the cells determines the clustering of adjacent cells in secondary memory.

Figure 3.22 shows four common labelings. Figure 3.22a corresponds to a row-wise enumeration of the cells [Sam90b]. Figure 3.22b shows the cell enumeration imposed by the *Peano curve*. Different authors have coined many names for this enumeration, including *Peano codes*, *ST_MortonNumbers*, *quad codes* [FB74], *N-trees* [Whi81], *locational codes* [AS83], or *z-values* [OM84]. Figure 3.22c shows the *Hilbert curve* [FR89, Jag90a], and Fig. 3.22d depicts *Gray ordering* [Fal85, Fal88], which is obtained by interleaving the Gray codes of the *x*- and *y*-coordinates in a bitwise manner. Gray codes of successive cells differ in exactly one bit.

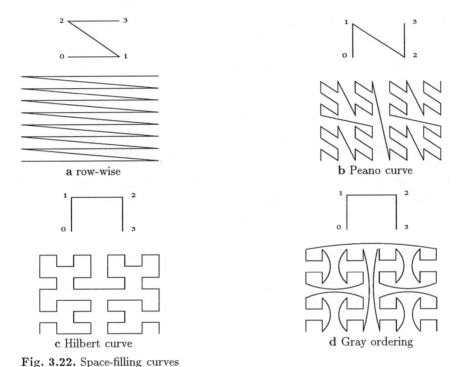

**Fig. 3.22.** Space-filling curves

Based on several experiments, Abel and Mark [AM90] conclude that the Peano and the Hilbert curve are most suitable as multidimensional access methods. Jagadish [Jag90a] and Faloutsos and Rong [FR91] all prefer the Hilbert curve among those two. Z-ordering [OM84], which is based on the Peano curve, is one of the few access methods that has found its way into commercial database products. In particular, Oracle has adapted and integrated the technique into Release 7.3 of its database system [Ora95].

An important advantage of all space-filling curves is that they are practically insensitive to the number of dimensions if the one-dimensional keys can be arbitrarily large. Everything is mapped into one-dimensional space, and one's favorite one-dimensional access method can be applied to manage the data. An obvious disadvantage of space-filling curves is that incompatible index partitions cannot be joined without recomputing the codes of at least one of the two indexes.

### 3.3.3 Spatial Access Methods

All multidimensional access methods presented in the previous section have been designed to handle sets of data points and to support spatial searches on them. None of those methods is directly applicable to databases containing objects with a spatial extension, such as polygons or polyhedra. In order to handle such extended objects, point access methods have been modified using one of the following four techniques:

1. Transformation (object mapping)
2. Overlapping regions (object bounding)
3. Clipping (object duplication)
4. Multiple layers.

A simpler version of this classification was first introduced by Seeger and Kriegel [SK88]. Later on, Kriegel et al. [KHH+92] added another dimension to this taxonomy: a spatial access method's *base type*, i.e., the spatial data type it supports primarily. Table 3.4 shows the resulting classification of spatial access methods. Note that most structures use the interval as a base type.

**Table 3.4.** Classification of spatial access methods [KHH+92]

| Technique | Base type | | | |
|---|---|---|---|---|
| | Grid cell | Interval (box) | Sphere | Polyhedron |
| Transfor-mation | zkdB+-tree, BANG file, hB-tree | All PAMs listed in Sect. 3.3.2 except BANG file and hB-tree | | P-tree |
| Overlapping regions | | R-tree, R*-tree, buddy tree with overlapping regions | sphere tree | KD2B-tree |
| Clipping | | EXCELL, R+-tree, buddy tree with clipping | | cell tree |
| Multiple layers | | multi-layer grid file, R-file | | |

**Transformation.** One-dimensional access methods and point access methods can often be used to manage spatially extended objects, provided the objects are first transformed into a different representation. There are essentially two options: one can transform each object either into a higher-dimensional point, or into a set of one-dimensional intervals via space-filling curves. A priori, both approaches are restricted to simple geometric shapes such as circles or rectangles. We discuss the two techniques in turn.

*Mapping to Higher-Dimensional Space [Hin85, SK88].* Simple geometric shapes can be represented as points in a higher-dimensional space. For example, it takes four real numbers to represent a (two-dimensional) rectangle in $E^2$. Those numbers may be interpreted as coordinates of a point in $E^4$. One possibility is to take the $x$- and $y$-coordinates of two diagonal corners *(endpoint transformation)*, another option is based on the centroid and two additional parameters for the extension of the object in $x$- and $y$-direction *(midpoint transformation)*. Any such transformation maps a database of rectangles onto a database of four-dimensional points, which can then be managed by one of the PAMs discussed in the previous section. Search operations can be expressed as point and region queries in this dual space.

If the original database contains more complex objects, they have to be approximated – e.g., by a rectangle or a sphere – before transformation. In this case, the point access method can only lead to a partial solution (Fig. 3.11).

Figure 3.23 shows the dual space equivalents of some common queries for the endpoint transformation. For presentation purposes, the figure shows a mapping from intervals in $E^1$ to points in $E^2$. A range query with the search range $[l, u]$ is mapped into a general region query in dual space. Any point in dual space that lies in the shaded area corresponds to an interval in original space that overlaps the search interval $[l, u]$, and vice versa. Enclosure and containment queries with the interval $[l, u]$ as argument also map into general region queries. A point query with $p$ as the (one-dimensional) query point maps into a range query for the endpoint transformation.

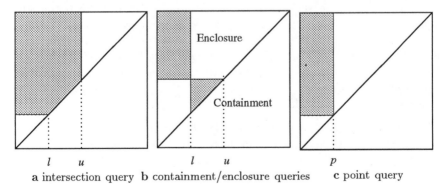

a intersection query    b containment/enclosure queries    c point query

**Fig. 3.23.** Search queries in dual space

Despite its conceptual elegance, this approach has several disadvantages. First, as the examples above already indicate, the formulation of point and range queries in dual space is usually more complicated than in the original space [NH87]. Finite search regions may map into infinite search regions in dual space, and some more complex queries involving spatial predicates may not be expressible at all any more [HSW89, Ore90a, PST93]. Second, depending on the chosen mapping, the distribution of points in dual space may be highly non-uniform even though the original data is uniformly distributed. With the endpoint transformation, for example, there are no image points below the main diagonal [FSR87]. Third, the images of two objects that are close in original space may be arbitrarily far apart from each other in dual space.

To overcome some of these problems, several authors have proposed special transformation and split strategies [HSW89, FR91, PST93]. A structure designed explicitly to be used in connection with the transformation technique is the LSD tree [HSW89]. Performance studies by Henrich and Six [HS92a] confirm the claim that the LSD tree adapts well to non-uniform distributions, which is of particular relevance in this context. It also contains a mechanism to avoid searching large empty query spaces, which may occur as a result of the transformation.

*Mapping to One-Dimensional Space.* Space-filling curves (cf. Sect. 3.3.2) are a very different type of transformation approach that seems to suffer less from these drawbacks. Space-filling curves can be used to represent extended objects by a list of grid cells or, equivalently, a list of one-dimensional intervals that define the position of the grid cells concerned. In other words, a complex spatial object is approximated by the union of *several one-dimensional intervals*. This should be seen in contrast to the technique described above where a complex object is approximated by *one higher-dimensional point*.

There are different variations of this basic concept, including *z-ordering* [OM84], the *Hilbert R-tree* [KF94], and the *UB-tree* [Bay96]. Z-ordering, for example, is based on the Peano curve. A simple algorithm to obtain the z-ordering representation of a given extended object can be described as follows. Starting from the (fixed) universe containing the data object, the space is split recursively into two subspaces of equal size by $(d - 1)$-dimensional hyperplanes. As in the k-d-tree, the splitting hyperplanes are iso-oriented, and their directions alternate in fixed order among the $d$ possibilities. The subdivision continues until one of the following three conditions holds:

1. The current subspace does not overlap the data object.
2. The current subspace is fully enclosed in the data object.
3. Some given level of accuracy has been reached.

The data object is thus represented by a set of cells, called *Peano regions* or *z-regions*. As discussed in Sect. 3.3.2, each such Peano region can be represented by a unique bit string (the *Peano code* or *z-value*). Using those bit strings, the cells can then be stored in a standard one-dimensional index, such as a $B^+$-tree. Orenstein calls this structure a $zkdB^+$-tree [Ore86].

Figure 3.24 shows a simple example. Figure 3.24a shows the polygon to be approximated, with the frame representing the universe. After several splits, starting with a vertical split line, we obtain Fig. 3.24b. Nine Peano regions of different shapes and sizes approximate the object. The labeling of each Peano region is shown in Fig. 3.24c. As an example consider the Peano region $\bar{z}$ in the lower left part of the given polygon. It lies to the left of the first vertical hyperplane and below the first horizontal hyperplane. Hence, the first two bits of the corresponding Peano code are 00. As we further partition the lower left quadrant, $\bar{z}$ lies on the left of the second vertical hyperplane, but above the second horizontal hyperplane. The Peano code accumulated so far is thus 0001. In the next round of decompositions, $\bar{z}$ lies to the right of the third vertical hyperplane and above the third horizontal hyperplane, resulting in two additional 1's. The complete Peano code corresponding to $\bar{z}$ is therefore 000111.

Figures 3.24b and 3.24c also give some bit strings along the coordinate axes, which describe only the splits orthogonal to the given axis. The string 01 on the $x$-axis, for example, describes the subspace that lies to the left of the first vertical split and to the right of the second vertical split. By bit-interleaving the bit strings that one finds when projecting a Peano region onto the coordinate axes, we obtain its Peano code. Note that if a Peano code $z_1$ is the prefix of some other Peano code $z_2$, the Peano region corresponding to $z_1$ *encloses* the Peano region corresponding to $z_2$. The Peano region corresponding to 00, for example, encloses the regions corresponding to 0001 and 000. This is an important observation, since it can be used for query processing [GR94].

As z-ordering is based on an underlying grid, the resulting set of Peano regions is usually only an approximation of the original object. The termination criterion depends on the accuracy or *granularity* (maximum number of bits) desired. More Peano regions obviously yield more accuracy, but they also increase the size and complexity of the approximation. As discussed by several authors [Ore89, Ore90c, Gae95b], there are two possibly conflicting objectives. First, the number of Peano regions to approximate the object should be small, since this results in fewer index entries. Second, the accuracy of the approximation should be high, since this reduces the expected number of false drops (i.e., objects that are paged in from secondary memory, only to find out that they do not satisfy the search predicate). By enhancing the z-ordering encoding with a single bit that reflects for each Peano region whether it is enclosed in the extended object or not, it is possible to improve the performance of z-ordering even further [Gae95a]. Figure 3.25 shows Peano regions for the running example.

**Overlapping Regions.** The key idea of the overlapping regions technique is to allow different data buckets in an access method to correspond to *mutually*

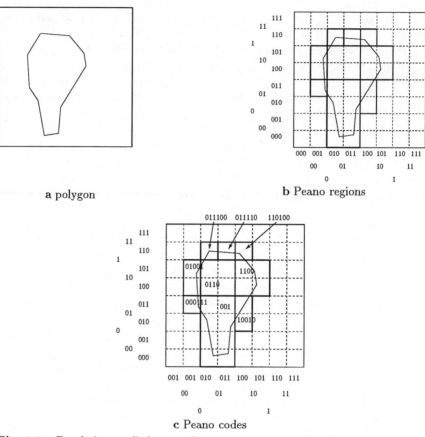

**a** polygon                    **b** Peano regions

**c** Peano codes

**Fig. 3.24.** Z-ordering applied to a polygon

*overlapping* subspaces. With this method we can assign any extended object directly and as a whole to a single bucket region. Consider, for instance, the k-d-B-tree for the running example, depicted in Fig. 3.18, and one of the polygons given in the scenario (Fig. 3.12), say $r_{10}$. $r_{10}$ overlaps two bucket regions, the one containing $p_{10}$, $c_1$, and $c_2$, and the other one containing $c_{10}$ and $p_9$. If we extend one of those regions to accommodate $r_{10}$, this polygon could be stored in the corresponding bucket. Note, however, that this extension inevitably leads to an overlap of regions.

Search algorithms can be applied almost unchanged. The only differences are due to the fact that the overlap may increase the number of search paths we have to follow. Even a point query may require the investigation of multiple search paths because there may be several subspaces at any index level that include the search point. For range and region queries, the average number of search paths increases as well.

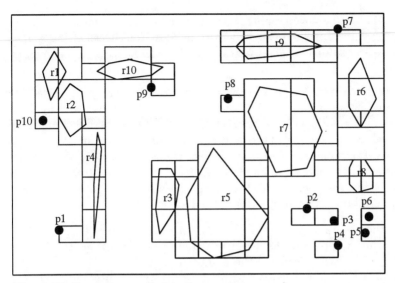

**Fig. 3.25.** Z-ordering applied to the running example

Hence, while functionality is not a problem when using overlapping regions, performance can be. This is particularly relevant when the spatial database contains objects whose size is large relative to the size of the universe. Typical examples are known from geographic applications where one has to represent objects of widely varying size (such as buildings and states) in the same spatial database. Each insertion of a new data object may increase the overlap and therefore the average number of search paths to be traversed per query. Eventually, the overlap between subspaces may become large enough to render the index ineffective because one ends up searching most of the index for a single point query. A well-known example where this degenerate behavior has been observed is the R-tree [Gut84, Gre89]. Several modifications have been presented to mitigate these problems, including a technique to minimize the overlap [RL85]; see Sect. 3.3.3 for a detailed discussion.

A minor problem with overlapping regions concerns ambiguities during insertion. If we insert a new object, we could in principle enlarge any subspace to accommodate it. To optimize performance, there exist several strategies [PSTW93]. For example, we could try to find the subspace that causes minimal additional overlap, or the one that requires the least enlargement. If it takes too long to compute the optimal strategy for every insertion, some heuristic may be used.

When a subspace needs to be split, one also tries to find a split that leads to minimal overall overlap. Several authors suggest heuristics for this problem [Gut84, Gre89, BKSS90].

In SAMs based on overlapping regions, the subspaces are mostly interval-shaped. Exceptions include two structures by Oosterom [Oos90]: the *sphere tree*, where the subspaces are spheres, and the *KD2B-tree* with polyhedral subspaces.

*The R-Tree [Gut84].* An R-tree corresponds to a hierarchy of nested $d$-dimensional intervals (boxes). Each node $\nu$ of the R-tree corresponds to a disk page and a $d$-dimensional interval $I^d(\nu)$. If $\nu$ is an interior node then the intervals corresponding to the descendants $\nu_i$ of $\nu$ are contained in $I^d(\nu)$. Intervals at the same tree level may overlap. If $\nu$ is a leaf node, $I^d(\nu)$ is the $d$-dimensional minimum bounding box (MBB) of the objects stored in $\nu$. For each object in turn, $\nu$ only stores its MBB and a reference to the complete object description. Other properties of the R-tree include:

- Every node contains between $m$ and $M$ entries unless it is the root. The lower bound $m$ prevents the degeneration of trees and ensures an efficient storage utilization. Whenever the number of a node's descendants drops below $m$, the node is deleted and its descendants are distributed among the sibling nodes (*tree condensation*). The upper bound $M$ can be derived from the fact that each tree node corresponds to exactly one disk page.
- The root node has at least two entries unless it is a leaf.
- The R-tree is height-balanced, i.e., all leaves are at the same level. The height of an R-tree is at most $\lceil \log_m(N) \rceil$ for $N$ index records ($N > 1$).

Searching in the R-tree is similar to the B-tree. At each index node $\nu$, all index entries are tested for whether they intersect the search interval $I_s$. We then visit all child nodes $\nu_i$ with $I^d(\nu_i) \cap I_s \neq \emptyset$. Due to the overlapping region paradigm, there may be several intervals $I^d(\nu_i)$ that satisfy the search predicate. Thus, there exists no non-trivial worst-case bound for the number of pages we have to visit. Figure 3.26 gives an R-tree for the running example. Remember that the $m_i$ denote the MBBs of the polygonal data objects $r_i$. A point query with search point $X$ results in two paths: $R_8 \to R_4 \to m_7$ and $R_7 \to R_3 \to m_5$.

Because the R-tree only manages MBBs, it cannot solve a given search problem completely unless, of course, the actual data objects are interval-shaped. Otherwise the result of an R-tree query is a set of candidate objects, whose actual spatial extent then has to be tested for intersection with the search space (cf. Fig. 3.11). This step, which may cause additional disk accesses and considerable computations, has not been taken into account in most published performance analyses [Gut84, Gre89].

To insert an object $o$, we insert the minimum bounding interval $I^d(o)$ and an object reference into the tree. In contrast to searching, we traverse only a single path from the root to the leaf. At each level we choose the child node $\nu$ whose corresponding interval $I^d(\nu)$ needs the least enlargement to enclose the data object's interval $I^d(o)$. If several intervals satisfy this criterion, Guttman proposes to select the descendant associated with the smallest (*d*-dimensional) interval. As a result, we insert the object only once, i.e., the object is not dispersed over several buckets. Once we have reached the leaf level, we try to insert the object. If this requires an enlargement of the corresponding bucket region, we adjust it appropriately and propagate the change upwards. If there is not enough space left in the leaf, we split it and distribute the entries among the old and the new page. Once again, we adjust each of the new intervals accordingly and propagate the split up the tree.

As for deletion, we first perform an exact match query for the object in question. If we find it in the tree, we delete it. If the deletion causes no underflow we check whether the bounding interval can be reduced in size. If so, we perform this adjustment and propagate it upwards. On the other hand, if the deletion causes node occupation to drop below *m*, we copy the node content into a temporary node and remove it from the index. We then propagate the node removal up the tree, which typically results in the adjustment of several bounding intervals. Afterwards we reinsert all orphaned entries of the temporary node. Alternatively, we can merge the orphaned entries with sibling entries. In both cases, one may again have to adjust bounding intervals further up the tree.

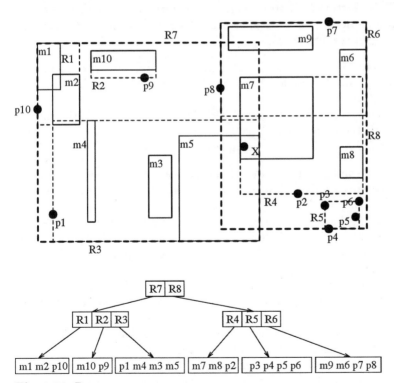

Fig. 3.26. R-tree

In his original paper, Guttman [Gut84] discusses various policies to minimize overlap during insertion. For node splitting, for example, Guttman suggests several algorithms, including a simpler one with linear time complexity and a more elaborate one with quadratic complexity. Later work by other researchers led to the development of more sophisticated policies. The *packed R-tree* [RL85], for example, computes an optimal partitioning of the universe and a corresponding minimal R-tree for a given scenario. However, it requires all data to be known a priori. Ng and Kameda [NK93, NK94] discuss how to support concurrency and recovery in R-trees.

*The R\*-Tree [BKSS90].* Based on a careful study of the R-tree behavior under different data distributions, Beckmann et al. [BKSS90] identified several weaknesses of the original algorithms. In particular, they confirmed the observation of Roussopoulos and Leifker [RL85] that the insertion phase is critical for search performance. The design of the *R\*-tree* therefore introduces a policy called *forced reinsert*: if a node overflows, they do not split it right away. Rather, they first remove $p$ entries from the node and reinsert them into the tree. The parameter $p$ may vary; Beckmann et al. suggest $p$ to be about 30% of the maximal number of entries per page.

Another issue investigated by Beckmann et al. concerns the node splitting policy. While Guttman's R-tree algorithms only tried to minimize the area that is covered by the bucket regions, the R\*-tree algorithms also take the following objectives into account:

– Overlap between bucket regions at the same tree level should be minimized. The less overlap, the smaller the probability that one has to follow multiple search paths.
– Region perimeters should be minimized. The best rectangle is the square, since this is the most compact rectangular representation.
– Storage utilization should be maximized.

The improved splitting algorithm of Beckmann et al. [BKSS90] is based on the plane-sweep paradigm [PS85]. In $d$ dimensions, its time complexity is $O(d \cdot n \cdot \log n)$ for a node with $n$ intervals.

In summary, the R\*-tree differs from the R-tree mainly in the insertion algorithm; deletion and searching are essentially unchanged. Beckmann et al. report performance improvements of up to 50% compared to the basic R-tree. Their implementation also shows that reinsertion may improve storage utilization. In broader comparisons, however, several authors [HS92b, GG96] found that the CPU time overhead of reinsertion can be substantial, especially for large page sizes; see Sect. 3.3.4 for further details.

One of the major insights behind the R\*-tree is that node splitting is critical for the overall performance of the access method. Since a naive (exhaustive) approach has time complexity $O(d \cdot 2^n)$ for $n$ given intervals, there is a need for efficient and optimal splitting policies. Becker and Güting [BG92] proposed a polynomial time algorithm that finds a balanced split, which also

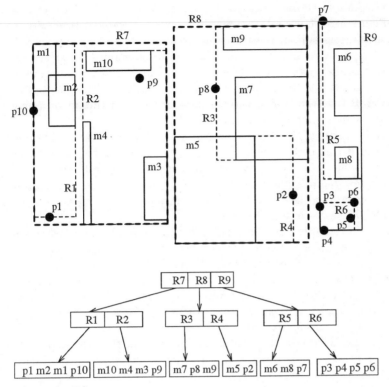

**Fig. 3.27.** R*-tree

optimizes one of several possible objective functions (e.g., minimum sum of areas or minimum sum of perimeters). They assume in their analysis that the intervals are presorted in some specific order.

*The P-Tree [Jag90b].* In many applications, intervals are not a good approximation of the data objects enclosed. The polyhedral or P-tree of Jagadish [Jag90b] aims to combine the flexibility of polygon-shaped containers with the simplicity of the R-tree. Jagadish first introduces a variable number $m$ of orientations in the $d$-dimensional universe, where $m > d$. For instance, in two dimensions ($d = 2$) we may have four orientations ($m = 4$): two parallel to the coordinate axes (i.e., iso-oriented), and two parallel to the two main diagonals. Objects are approximated by minimum bounding polytopes whose faces are parallel to these $m$ orientations. Clearly, the quality of the approximations is positively correlated with $m$. We can now map the original space into an $m$-dimensional *orientation space*, such that each ($d$-dimensional) approximating polytope $P^d$ turns into an $m$-dimensional interval $I^m$. Any point inside (outside) $P^d$ maps onto a point inside (outside) $I^m$, while the opposite is not necessarily true. To maintain the $m$-dimensional intervals, a large selection of SAMs is available. Jagadish suggests the R- or R+-tree for this purpose. Figure 3.28 shows a P-tree for the running example with $m = 4$.

An interesting feature of the P-tree is the ability to add hyperplanes to the attribute space dynamically without having to reorganize the structure. By projecting the new intervals of the extended orientation space onto the old orientation space, it is still possible to use the old structure. Consequently, we can obtain an R-tree from a higher-dimensional P-tree structure by dropping all hyperplanes that are not iso-oriented.

The interior nodes of the P-tree represent a hierarchy of nested polytopes, similar to the R-tree or the cell tree. Polytopes corresponding to different nodes at the same tree level may overlap. For search operations we first compute the minimum bounding polytope of the search region and map it onto an $m$-dimensional interval. The search efficiency then depends on the chosen PAM. The same applies for deletion.

While the introduction of additional hyperplanes results in a better approximation, it increases the size of the entries, thus reducing the fan-out of the interior nodes. Experiments reported by Jagadish [Jag90b] suggest that a 10-dimensional orientation space ($m = 10$) is a good choice for storing two-dimensional lines ($d = 2$) with arbitrary orientation. This needs to be compared to a simple MBB approach. Although the latter technique may sometimes render poor approximations, the representation requires only four numbers per line. Storing a 10-dimensional interval, on the other hand, requires 20 numbers, i.e., five times as much. Another drawback of the P-tree is the fixed orientation of the hyperplanes.

**Clipping.** Clipping-based schemes do not allow any overlaps between bucket regions; they have to be *mutually disjoint*. A typical example is the R+-tree [SSH86, SRF87], a variant of the R-tree that allows no overlap between regions corresponding to nodes at the same tree level. As a result, point queries follow a single path starting at the root, which means efficient searches.

The main problems with clipping-based approaches relate to the insertion and deletion of data objects. During insertion, any data object that spans more than one bucket region has to be subdivided along the partitioning hyperplanes. Eventually, several bucket entries have to be created for the same object. Each bucket stores either the geometric description of the complete object *(object duplication)*, or the geometric description of the corresponding fragment with an object reference. In any case, data about the object is dispersed among several data pages *(spanning property)*. The resulting *redundancy* [Ore89, Ore90c, GG96] may cause not only an increase in the average search time, but also an increase in the frequency of bucket overflows.

A second problem applies to clipping-based access methods that do not partition the complete data space a priori. In that case, the insertion of a new data object may lead to the enlargement of several bucket regions. Whenever the object (or a fragment thereof) is passed down to a bucket (or, in the case of a tree structure, an interior node) whose region does not cover it, the

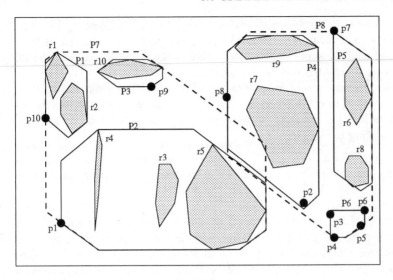

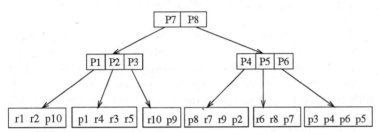

Fig. 3.28. P-tree

region has to be extended. In some cases, such an enlargement is not possible without getting an overlap with other bucket regions; this is sometimes called the *deadlock* problem of clipping. Because overlap is not allowed, we have to redesign the region structure, which can become very complicated. It may in particular cause further bucket overflows and insertions, which can lead to a chain reaction and, in the worst case, a complete breakdown of the structure [GB91]. Access methods partitioning the complete data space do not suffer from this problem.

A final problem concerns the splitting of buckets. There may be situations where a bucket (and its corresponding region) has to be split but there is no splitting hyperplane that splits none (or only a few) of the objects in that bucket. The split may then trigger other splits, which may become problematic with increasing size of the database. The more objects are inserted, the higher the probability of splits and the smaller the average size of the bucket regions. New objects are therefore split into a larger number of smaller fragments, which may in the worst case once again lead to a chain reaction.

To alleviate these problems, Günther et al. [GN91, GG96] suggest storing large objects (which are more likely to be split into a large number of fragments) in special buckets called *oversize shelves*. Oversize shelves are special disk pages that are attached to the interior nodes of the tree. They accommodate objects that would have been split into too many fragments if they had been inserted into the structure. The authors propose a dynamically adjusting threshold for choosing between placing a new object on an oversize shelf or inserting it regularly.

*The $R^+$-Tree [SSH86, SRF87].* To overcome the problems associated with overlapping regions in the R-tree, Sellis et al. introduced an access method called the $R^+$-tree. Unlike the R-tree, the $R^+$-tree uses clipping, i.e., there is no overlap between index intervals $I^d$ at the same tree level. Objects that intersect more than one index interval have to be stored on several different pages. As a result of this policy, point searches in $R^+$-trees correspond to *single-path* tree traversals from the root to one of the leaves. They therefore tend to be faster than the corresponding R-tree operation. Range searches will usually lead to the traversal of multiple paths in both structures.

When inserting a new object $o$, we may have to follow multiple paths, depending on the number of intersections of the MBB $I^d(o)$ with index intervals. During the tree traversal, $I^d(o)$ may be split into $n$ disjoint fragments $I_i^d(o)$ ($\bigcup_{i=1}^n I_i^d(o) = I^d(o)$). Each fragment is then placed in a different leaf node $\nu_i$. Provided that there is enough space, the insertion is straightforward. If the bounding interval $I^d(o)$ overlaps space that has not yet been covered, we have to enlarge the intervals corresponding to one or more leaf nodes. Each of these enlargements may require a considerable effort because overlaps have to be avoided. In some rare cases, it may not be possible to increase the current intervals in such a way that they cover the new object without some mutual overlap [Gün88, Ooi90]. In case of such a deadlock, some data intervals have to be split and reinserted into the tree.

If a leaf node overflows, it has to be split. Node splittings work similarly as in the case of the R-tree. An important difference, however, is that splits may propagate not only up the tree, but also down the tree. The resulting *forced split* of the nodes below may lead to several complications, including further fragmentation of the data intervals; see for example the rectangles $m_5$ and $m_8$ in Fig. 3.29.

For deletion, we first locate all the data nodes where fragments of the object are stored and remove them. If storage utilization drops below a given threshold, we try to merge the affected node with its siblings or to reorganize the tree. This is not always possible, which is the reason why the $R^+$-tree cannot guarantee a minimum space utilization.

*The Cell Tree [Gün88, Gün89].* The main goal during the design of the cell tree was to facilitate searches on data objects of arbitrary shapes, i.e., especially on data objects that are not intervals themselves. The cell tree uses clipping to manage large spatial databases that may contain polygons or

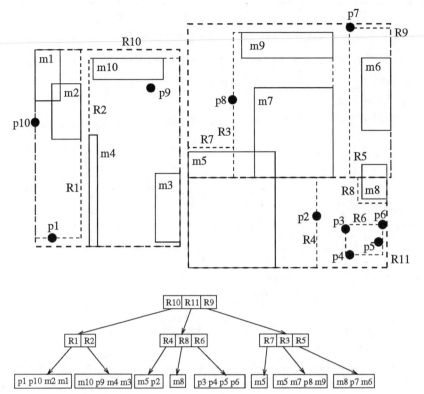

**Fig. 3.29.** R$^+$-tree

higher-dimensional polyhedra. It corresponds to a decomposition of the universe into disjoint convex subspaces. The interior nodes correspond to a hierarchy of nested polytopes and each leaf node corresponds to one of the subspaces (Fig. 3.30). Each tree node is stored on one disk page.

To avoid some of the disadvantages resulting from clipping, the convex polyhedra are restricted to be subspaces of a BSP (Binary Space Partitioning). Therefore we can view the cell tree as a combination of a BSP tree and an R$^+$-tree, or as a BSP tree mapped on paged secondary memory. In order to minimize the number of disk accesses that occur during a search operation, the leaf nodes of a cell tree contain all the information required for answering a given search query; we have to load no pages other than those containing relevant data. This is an important advantage of the cell tree over the R-tree and related structures.

Before inserting a non-convex object, we decompose it into a number of convex *components* whose union is the original object. The components do not have to be mutually disjoint. All components are assigned the same object identifier and inserted into the cell tree one by one. Due to clipping, we may

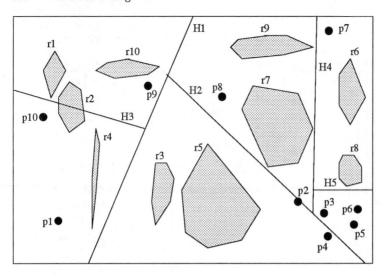

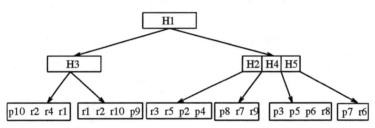

**Fig. 3.30.** Cell tree

have to subdivide each component into several *cells* during insertion, because
it overlaps more than one subspace. Each cell is stored in one leaf node of
the cell tree. If an insertion causes a disk page to overflow, we have to split
the corresponding subspace and cell tree node and distribute its descendants
among the two resulting nodes. Each split may propagate up the tree.

For point searches, we start at the root of the tree. Using the underlying
BSP partitioning, we identify the subspace that includes the search point
and continue the search in the corresponding subtree. This step is repeated
recursively until we reach a leaf node, where we examine all cells whether
they contain the search point. The solution consists of those objects that
contain at least one of the cells that qualify. A similar algorithm exists for
range searches. A performance evaluation of the cell tree [GB91] shows that
it is competitive with other popular spatial access methods.

Figure 3.30 shows our running example with five partitioning hyperplanes
*H1* through *H5*, each of them stored in the interior nodes. Even though the
partitioning by means of the BSP tree offers more flexibility compared to

rectilinear hyperplanes, it may be inevitable to clip objects. In Fig. 3.30, we had to split $r_2$ and insert the resulting cells into two pages.

Like all structures based on clipping, the cell tree has to cope with the fragmentation of the space, which becomes increasingly problematic as more objects are inserted into the tree. After some time, most new objects will be split into fragments during insertion. Oversize shelves [GN91] help to avoid the negative effects resulting from this fragmentation. Performance results by Günther and Gaede [GG96] show substantial improvements compared to the cell tree without oversize shelves.

**Multiple Layers.** The multiple layer technique can be regarded as a variant of the overlapping regions approach, because data regions of different layers may overlap. However, there are several important differences. First, the layers are organized in a hierarchy. Second, each layer partitions the complete universe in a different way. Third, data regions within a layer are disjoint, i.e., they do not overlap. Fourth, the data regions do not adapt to the spatial extensions of the corresponding data objects.

To get a better understanding of the multi-layer technique, we shall discuss how to insert an extended object. We start by trying to find the lowest layer in the hierarchy whose hyperplanes do not split the new object. If there is such a layer, we insert the object into the corresponding data page. If the insertion causes no page to overflow, we are done. Otherwise, we must split the data region by introducing a new hyperplane and distributing the entries accordingly. Objects intersecting the hyperplane have to be moved to a higher layer or an overflow page. As the database becomes populated, the data space of the lower layers becomes more and more fragmented. As a result, large objects keep accumulating on higher layers of the hierarchy or even worse, it is no longer possible to insert objects without intersecting existing hyperplanes.

The multi-layer approach seems to offer one advantage compared to the overlapping regions technique: a possibly higher selectivity during searching due to the restricted overlap of the different layers. However, there are also several disadvantages: First, the multi-layer approach suffers from fragmentation, which may render the technique inefficient for some data distributions. Second, certain queries require the inspection of all existing layers. Third, it is not clear how to cluster objects that are spatially close to each other but in different layers. Fourth, there is some ambiguity in which layer to place the object.

*The Multi-Layer Grid File [SW88].* The multi-layer grid file is yet another variant of the grid file capable of handling extended objects. It consists of an ordered sequence of grid layers. Each of these layers corresponds to a separate grid file with freely positionable splitting hyperplanes that covers the whole universe. A new object is inserted into the first grid file in the sequence that does not imply any clipping of the object. If one of the grid files is extended by adding another splitting hyperplane, those objects that would be split have to be moved to another layer. Figure 3.31 illustrates a multi-layer grid file with two layers for the running example.

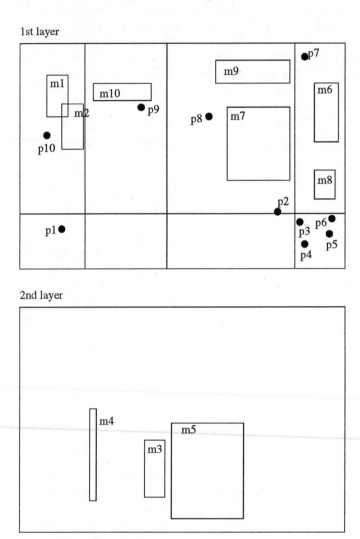

**Fig. 3.31.** Multi-layer grid file

In the multi-layer grid file, the size of the bucket regions typically increases within the sequence, i.e., larger objects are more likely to find their final location in later layers. If a new object cannot be stored in any of the current layers without clipping, a new layer has to be allocated. An alternative is to allow clipping only for the last layer. Six and Widmayer claim that $d + 1$

layers are sufficient to store a set of $d$-dimensional intervals without clipping if the hyperplanes are cleverly chosen.

For an exact match query, we can easily determine from the scales which grid file in the sequence is supposed to hold the search interval. Other search queries, in particular point and range queries, are answered by traversing the sequence of layers and by performing a corresponding search on each grid file. The performance results reported by Six and Widmayer [SW88] suggest that the multi-layer grid file is superior to the conventional grid file, using clipping to handle extended objects. Possible disadvantages of the multi-layer grid file include low storage utilization and expensive directory maintenance.

To overcome some of these problems, Hutflesz et al. [HSW90] later proposed an alternative structure for managing sets of rectangles, called the *R-file*. To avoid the low storage utilization of the multi-layer grid file, the R-file uses a single directory. The universe is partitioned as for the BANG file: splitting hyperplanes cut the universe recursively into equal parts, and z-ordering is used to encode the resulting bucket regions. In contrast to the BANG file, however, there are no excisions. Bucket regions may overlap, and there is no clipping. Each data interval is stored in the bucket with the smallest region that contains it entirely; overflow pages may be necessary in some cases.

### 3.3.4 Comparative Studies

Greene [Gre89] compares the search performance of the R-tree, the k-d-B-tree, and the $R^+$-tree for 10 000 uniformly distributed rectangles of varying size. Query parameters include the size of the query rectangles and the page size. Greene's study shows that the k-d-B-tree can never really compete with the two R-tree variants. On the other hand, there is not much difference between the $R^+$-tree and the R-tree, even though the former is significantly more difficult to code. As expected, the $R^+$-tree performs better when there is less overlap between the data rectangles.

Kriegel et al. [KSSS90] present an extensive experimental study of access method performance for a variety of spatial distributions. The study involves four point access methods: the hB-tree, the BANG file, the two-level grid file, and the buddy tree. According to the benchmarks, the buddy tree and, to some degree, the BANG file outperform all other structures. The reported results show in an impressive way how the performance of the studied access methods varies with different data distributions and query range sizes. For clustered data and a range query with a volume of 10% of the universe, for example, there is almost no performance difference between the buddy tree and the BANG file, whereas for a volume of 0.1% the buddy tree performs about twice as fast.

Seeger [See91] studied the relative performance of clipping, overlapping regions, and transformation techniques, implemented on top of the buddy tree. He also included the two-level grid file and the $R^*$-tree in the comparison. The buddy tree with clipping and the grid file failed completely for certain distributions, since they produced unmanageably large files. The transformation technique supports fast insertions at the expense of low storage utilization. The R*-tree, on the other hand, requires fairly long insertion times, but offers good storage utilization. For intersection and containment queries, the buddy tree combined with overlapping regions is superior to the buddy tree with transformation. The performance advantage of the overlapping regions technique decreases for larger query regions, even though the buddy tree with transformation never outperforms the buddy tree with overlapping regions. When the data set contains uniformly distributed rectangles of varying size, the buddy tree with clipping outperforms the other techniques for intersection and enclosure queries. For some queries the buddy tree with overlapping performs slightly better than the R*-tree.

Günther and Gaede [GG96] compare the original cell tree [Gün89] with the cell tree with oversize shelves [GN91], the R*-tree [BKSS90], and the hB-tree [LS89a] for some real cartographic data. There is a slight performance advantage of the cell tree with oversize shelves compared to the R*-tree and the hB-tree, but a major difference compared with the original cell tree. An earlier comparison using artificially generated data can be found in [Gün92].

Hoel and Samet [HS92b] compare the performance of the PMR-quadtree [NS87], the R*-tree, and the R+-tree for indexing line segments. The R+-tree shows the best insertion performance, whereas the R*-tree occupies the least space. However, the insertion behavior of the R+-tree depends heavily on the page size as opposed to the PMR-quadtree. The performance of all compared structures is about the same, even though the PMR-quadtree shows some slight performance benefits. Although the R*-tree is more compact than the other structures, its search performance is not as good as that of the R+-tree for line segments. Unfortunately, Hoel and Samet do not report the overall performance times for the different queries.

Peloux et al. [PdSMS94] carried out a similar performance comparison of two quadtree variants, a variant of an R+-tree, and the R*-tree. What makes their study different is that all structures have been implemented on top of a commercial object-oriented system using the application programming interface (API). A further difference to Hoel and Samet [HS92b] is that Peloux et al. used polygons rather than line segments as test data. Furthermore, they report the various times for index traversal, loading polygons, etc. Besides showing that the R+-tree and a quadtree variant based on Hierarchical EXCELL [Tam83] outperform the R*-tree for point queries, they demonstrate clearly that the database system must provide some means for physical clustering. Otherwise reading a single index page may induce several page faults.

Further experimental studies of multidimensional access methods can be found in [FB90, AM90, BKSS90, HSW90, Jag90a, Ooi90, Oos90, SG90, GB91, HS92a, HWZ92, KS91, KF92, BK94, KF94].

As the preceding discussion shows, although numerous experimental studies exist, they are hardly comparable. Theoretical studies may bring some more objectivity to this discussion. The problem with such studies is that they are usually very hard to perform if one wants to stick to realistic modeling assumptions. For that reason, there are only a few theoretical results on the comparison of multidimensional access methods.

Regnier [Reg85] and Becker [Bec92] investigated the grid file and some of its variants. The most complete theoretical analysis of range trees can be found in [OSBvK90, SO90]. Günther and Gaede [GG96] present a theoretical analysis of the cell tree. Recent analyses show that the theory of fractals seems to be particularly suitable for modeling the behavior of SAMs if the data distribution is non-uniform [FK94, BF95, FG96].

Some more analytical work exists on the R-tree and related methods. A comparison of the R-tree and the $R^+$-tree has been published by Faloutsos et al. [FSR87]. Pagel et al. [PSTW93] present an interesting probabilistic model of window query performance for the comparison of different access methods independently of implementation details. Among other things, their model reveals the importance of the perimeter as a criterion for node splitting, which has been intuitively anticipated by the inventors of the $R^*$-tree [BKSS90]. The central formula of Pagel et al. for computing the number of disk accesses in an R-tree was found independently by Kamel and Faloutsos [KF93]. The same authors [FK94] later refined this formula by using properties of the data set. Theodoridis and Sellis [TS96] propose a theoretical model to determine the number of disk accesses in an R-tree that only requires two parameters: the amount of data and the density in the data space. Their model also extends to non-uniform distributions.

In pursuit of an implementation-independent comparison criterion for access methods, Pagel et al. [PSW95] suggest using the degree of clustering. As a lower bound they assume the optimal clustering of the static situation, i.e., if the complete data set has been exposed beforehand. Incidentally, the significance of clustering for access methods has been demonstrated in numerous empirical investigations as well [Jag90a, KF93, BK94, Kum94, NH94].

### 3.3.5 Outlook

As we have demonstrated, research in spatial database systems has resulted in a multitude of multidimensional access methods. Even for experts it is becoming more and more difficult to recognize their merits and faults, since every new method seems to claim superiority to at least one access method that has been published previously. In this section we have not tried to resolve this problem but rather to give an overview of the pros and cons of a variety of structures. It will come as no surprise to the reader that at present no access method has proven itself to be superior to all its competitors in whatever sense. Even if one benchmark declares one structure to be the clear winner, another benchmark may prove the same structure to be inferior.

But why are such comparisons so difficult? Because there are so many different criteria for defining optimality, and so many parameters that determine performance. Both the time and space efficiency of an access method strongly depend on the data to be processed and the queries to be answered. An access method that performs reasonably well for iso-oriented rectangles may fail for arbitrarily oriented lines. Strongly correlated data may render an otherwise fast access method irrelevant for any practical application. An index that has been optimized for point queries may be highly inefficient for arbitrary region queries. Large numbers of insertions and deletions may deteriorate a structure that is efficient in a more static environment.

Initiatives to set up standardized testbeds for benchmarking and comparing access methods under different conditions are important steps in the right direction [KSSS90, GOP+97] But note that clever programming can often make up for inherent deficiencies of a structure (and vice versa). Other factors of unpredictable impact are the programming language used, the hardware, buffer size, page size, data set, etc. Hence, it is far from easy to compare or rank different access methods. Experimental benchmarks need to be studied with care and can only be a first indicator for usability.

Another interesting direction for future research consists in recognizing common features of different access methods and using them to build configurable methods in a way that leads to modular and reusable implementations. The *Generalized Search Tree (GiST)* of Hellerstein et al. [HNP95] is such a generic method. A GiST is a balanced tree of variable fanout between $kM$ and $M$ ($2/M \leq k \leq 1/2$), with the exception of the root node, which may have fanout between $2$ and $M$. It thereby unifies disparate structures such as B$^+$-trees and R-trees and supports an extensible set of queries and data types.

The *BV-Tree* [Fre95] represents an attempt to solve the $d$-dimensional B-tree problem, i.e., to find a generic generalization of the B-tree to higher dimensions. The BV-tree is not meant to be a concrete access method. It represents a conceptual framework that can be applied to a variety of existing access methods, including the BANG file or the hB-tree. Freeston's proposal is based on the conjecture that one can maintain the major strengths of the B-tree in higher dimensions, provided one relaxes the strict requirements concerning tree balance and storage utilization. The BV-tree is not completely balanced. Furthermore, while the B-tree guarantees a worst-case storage utilization of 50%, Freeston argues that such a comparatively high storage utilization cannot be ensured for higher dimensions for topological reasons. However, the BV-tree manages to achieve the 33% lower bound suggested by Lomet and Salzberg [LS89a].

When it comes to technology transfer, i.e. to the use of access methods in commercial products, most vendors resort to structures that are easy to understand and implement. Quadtrees in SICAD [Sie98] and Smallworld GIS [ND97], R-trees in the relational database system Informix [Inf97], and z-ordering in Oracle [Ora95] are typical examples. Performance seems to be of secondary importance for the selection, which comes as no surprise given the relatively small differences among methods in virtually all published analyses. The tendency is rather to take a structure that is simple and robust, and to optimize its performance by highly tuned implementations and tight integration with other system components.

Nevertheless, the implementation and experimental evaluation of access methods is essential, as it often reveals deficiencies and problems that are not obvious from the design or a theoretical model. To make such comparative evaluations both easier to perform and easier to verify, it is essential to provide platform-independent access to the implementations of a broad variety of access methods. Some WWW-based approaches may provide the right technological base for such a paradigm change [GOP+97, GMS+97]. Once every published paper includes a URL, i.e., an Internet address that points to an implementation, possibly with a standardized user interface, transparency will increase substantially. Until then, most users will have to rely on general wisdom and their own experiments to select an access method that provides the best fit for their current application.

## 3.4 Object-Oriented Techniques

An object-oriented database management system (OODBMS) is a DBMS with an object-oriented data model [Dit86, ABD+89, Dit90]. On the one hand, this simple definition entails the full functionality of a DBMS: persistence, secondary storage management, concurrency control, recovery, ad hoc query facility, and possibly data distribution and integrity constraints. We have already discussed the significance of these features for environmental data management. The second part of the definition refers to an *object-oriented data model*. This implies, among other things: object identity, classes and inheritance, complex object support, and user-defined data types. In geographic and environmental data management, these features are of varying significance. In this section, which is based on a survey article by Günther and Lamberts [GL94], we explain those features and discuss their relevance for environmental information systems. The syntax used in the examples is based on the syntax of the object-oriented database system $O_2$ [Deu91, BDK92].

### 3.4.1 Object Identity

In an OODBMS the existence of an object is independent of its value. In contrast to the philosophy of the relational model, it is possible for two objects to have equal values and nevertheless to be identified unambiguously. Each object is labeled by some unique object identifier (OID) created by the system in order to guarantee systemwide, if not worldwide, uniqueness. It is not visible to the user and does not change during the lifetime of an object.

OIDs are an important concept for geographic and environmental information systems because they can be used to implement complex objects and shared subobjects, and to assign *multiple* representations (e.g., different scales, or raster versus vector) to a single object. Moreover, one can easily distinguish data objects that have been generated and managed at different locations, which is of particular importance in many environmental applications.

### 3.4.2 Classes and Inheritance

Object-oriented systems use the concept of the abstract data type (Sect. 3.2.3) for defining the structure of object *classes*. A class is a collection of objects of the same (abstract) data type. They thus all have the same structure. Classes support the two basic concepts underlying abstract data types: *abstraction* and *encapsulation*. An object can only be accessed through the operators defined on its class, i.e., it is only characterized through its behavior. The user is prevented from applying unsuitable operators to the object, and its internal representation is hidden.

Operators (methods) and attributes are attached to a class, which means that they are valid for all objects that belong to it. Classes may form an inheritance hierarchy. This means that all attributes and methods of a class apply to its subclasses as well, unless they are explicitly overwritten.

Figure 3.32 gives an example of an inheritance hierarchy. Rectangles symbolize classes, and ovals represent attributes. The superclass *EnvObject* has two attributes: *Name* and the spatial attribute *Shape*. *EnvObject* has four subclasses: *Biotope*, *Road*, *River*, and *AdminUnit*. Both *AdminUnit* and *Biotope* add some attributes to the ones inherited from *EnvObject*. *AdminUnit* has two subclasses in turn: *City* and *District*. *City* extends the class definition some more, thus resulting in five attributes: *Name*, *Shape*, *Population*, *Mayor*, and *Districts*. *Districts* is a set-valued attribute: it contains all districts that make up the city in question. All five attributes are forwarded in turn to the descendants of *City*, i.e., to *UnivTown* (with the additional attribute *NoStudents*) and to *Spa* (with the additional attribute *Tax*). *Biotope*, on the other hand, adds four attributes to the ones inherited from *EnvObject*: *BiotopeID* is a special application-specific identifier, *ProtStatus* defines the current and intended protection status, *EndType* lists the various types of endangerment, and *ProtSpecies* lists the protected species living in the given biotope.

There may be classes that do not contain any objects because they only serve as a container for a number of subclasses. This allows one to factorize those attributes and methods that the subclasses have in common. Depending on the application, *EnvObject* could be defined as such an *abstract class*. In this case, any environmental object would have to be either a biotope, a road, a river, or an administrative unit.

An object may belong to multiple classes. For example, the city of *Stuttgart* is a university town and a spa, and therefore has attributes *No-*

*Students* and *Tax* (in addition to the attributes attached to *City*). *Berkeley* is only a university town, and therefore has just the attribute *NoStudents* (in addition to the attributes attached to *City*) and not the attribute *Tax*. The opposite is true for *Baden-Baden*.

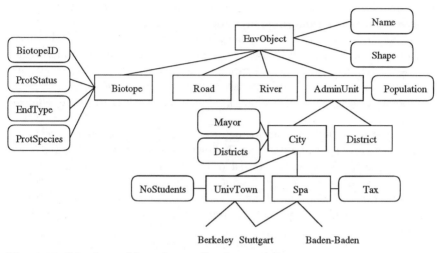

**Fig. 3.32.** Inheritance hierarchy: application modeling

In some object-oriented systems one cannot assign objects to multiple classes, but classes can have multiple ancestor classes. If this is the case, one would need to introduce an additional class to represent the situation in Fig. 3.32. This class would contain the object *Stuttgart* and would have both *UnivTown* and *Spa* as superclasses. Occasionally, it may be necessary to define schemes to resolve possible conflicts between the attributes and methods inherited from different sources.

In environmental information systems, there are many possible uses for inheritance hierarchies. For example, the extensive classification schemes in the natural sciences can easily be translated into inheritance hierarchies. Environmental object taxonomies, such as the ones presented in Sect. 2.1, can also be represented by inheritance hierarchies in a natural manner.

Spatial object hierarchies, on the other hand, tend to be rather shallow. Moreover, the inheritance of methods from the general to the special case is usually not very efficient. Consider, for example, the class hierarchy *Polygon – Convex Polygon – Rectangle – Square*. For the computation of common spatial operators such as *area* or *intersection*, there are specialized algorithms for each of these classes. It is hardly efficient to use the same area computation algorithm for general polygons and for squares – which would be typical for the inheritance paradigm. Moreover, inheritance hierarchies are most useful if the structural features of a subclass (i.e., its attributes and its methods) are a *superset* of the structural features of its superclass. For example, a class *Bird* inherits all attributes and methods of its superclass *Animal* and adds some more that are specific to birds. With semantic hierarchies of spatial object classes, however, it is often exactly the opposite. Squares, for example, form a subclass of the class of rectangles. The attributes for describing a square (e.g., coordinates of two vertices), however, form a *subset* – rather than a superset – of the attributes required for a rectangle.

These considerations can be generalized to other classes of spatial objects and operators. As a result, many researchers tend to avoid class hierarchies for the modeling of spatial data types and algorithms [Sch91]. Other authors, however, recommend a more extensive utilization of inheritance [SV92]. At this point it is not clear yet which modeling approach is the more efficient one.

In order to model the situation depicted in Fig. 3.32, one first needs to set up a collection of spatial data types. Following the discussion in Section 3.2.1, a possible set of types would consist of the classes *Point*, *Line* and *Polygon* with their common superclass *Spatial*, and the corresponding plural classes *PointSet*, *LineSet*, *PolygonSet*, and *SpatialSet*. The resulting inheritance hierarchy is sketched in Fig. 3.33. A similar approach has been suggested in [SV92].

Rather than introducing special classes for the plural version, one could also just use a set constructor in the type definition whenever necessary (cf. Sect. 3.4.3). For example, shapes could be defined to have the type *set(Polygon)* instead of *PolygonSet*. The disadvantage of this solution concerns the way methods can be defined. If there are special classes for the plural case, one can attach the relevant methods (e.g., for the intersection of two sets of polygons) directly to the corresponding plural class. Otherwise, one would define those methods for each class where the set construct occurs. In the given example, that would mean that the class *City* may include a special method to compute the intersection of its shape with the shape of another city. A similar method would be required for polygon-shaped districts, and so on.

Figures 3.35 and 3.34 give a partial definition of the example in $O_2$-style syntax. Note that unless one inserts the keyword *public*, the class structures are hidden from the user. Note also that the *Shape* attribute of *EnvObject* is of type *SpatialSet*. This type is overwritten by the more specific types *PolygonSet* and *LineSet* for the classes *Biotope* and *AdminUnit*, respectively. The *Shape* attribute of the class *River* is not defined explicitly. It is inherited from the class *EnvObject* and therefore of the less specific type *SpatialSet*. This way the shape of a river can consist of a collection of lines and polygons. The class *Biotope* refers to a class *Species*, whose definition is delayed until Sect. 3.4.4.

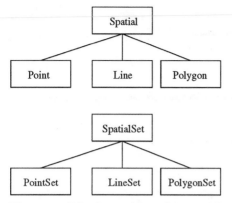

**Fig. 3.33.** Inheritance hierarchy: spatial modeling

### 3.4.3 Structural Object-Orientation

Real-world geographic and environmental objects are often structured in a hierarchical manner. Federal republics like the United States or Germany, for example, consist of several states, which are split up into counties, which are in turn divided into districts, and so on. Cities may consist of districts, streets, blocks, buildings, etc. Hardly any commercial GIS or conventional DBMS can provide sufficient support for these *complex* (also called *composite* or *structured*) objects. Instead the user typically has to split up objects into components until they are atomic, i.e., until the base type level of the GIS or DBMS is reached.

In a relational DBMS, the components may have to be stored in different relations. This kind of fragmentation considerably complicates the modeling and may have a negative impact on the system's performance. The user's ideas are not reflected appropriately in the underlying data structures, which complicates the interpretation of modeled objects. It is not possible to provide for spatial clustering, i.e., to make sure that component parts of the same larger structure are stored close together on the disk.

Object-oriented techniques provide solutions to some of these problems. By analogy to corresponding features in many typed programming languages, most object-oriented systems allow the user to build complex type structures. A data model is called *structurally object-oriented* if it supports the construction of complex objects. This means in particular that attributes of a tuple do not have to be atomic (as in the relational model), but may be composite in turn. To build application-specific complex object structures from atomic types (*integer, character*, etc.), users are offered special *type constructors*, such as *tuple, set,* or *list*.

If a user creates an object of some complex type, the system checks the correctness of the object structure automatically. Moreover, many OODBMS provide specific operators for complex objects, such as duplication and dele-

```
class EnvObject
     public type tuple(Name: string,
                       Shape: SpatialSet)
end;

class Biotope inherit EnvObject
     public type tuple(BiotopeID: integer,
                       ProtStatus: tuple(Present: string,
                                         Suggestion: string),
                       EndType: set(string),
                       ProtSpecies: set(Species),
                       Shape: PolygonSet)
end;

class Road inherit EnvObject
     public type tuple(Shape: LineSet)
end;

class River inherit EnvObject
end;

class AdminUnit inherit EnvObject
     public type tuple(Population: integer,
                       Shape: PolygonSet)
end;

class City inherit AdminUnit
     type tuple(Districts: set(District),
                Mayor: string)
end;

class District inherit AdminUnit
end;

class UnivTown inherit City
     type tuple(NoStudents: integer)
end;

class Spa inherit City
     type tuple(Tax: integer)
end;
```

**Fig. 3.34.** O$_2$-style representation of the example: application modeling

```
class Spatial
end;

class Point inherit Spatial
      public type tuple(X: integer,
                        Y: integer)
end;

class Line inherit Spatial
      public type list(Point)
end;

class Polygon inherit Spatial
      public type list(Point)
end;

class SpatialSet
      public type set(Spatial)
end;

class PointSet inherit SpatialSet
      public type set(Point)
end;

class LineSet inherit SpatialSet
      public type set(Line)
end;

class PolygonSet inherit SpatialSet
      public type set(Polygon)
end;
```

**Fig. 3.35.** O$_2$-style representation of the example: spatial modeling

tion of objects with their substructures, or navigation through an object structure.

Structural object-orientation can be used in a natural way to model *built-in objects* (i.e., objects embedded in other objects) or *shared objects*, which are included in several other objects simultaneously. By granting access to the complex type to other users, redundancies can be avoided and consistency can be enforced. This last point would be more difficult if one chose to represent a complex object by means of an ADT. It would then be more difficult to share components between different higher-level objects. As a result, components would often be represented redundantly.

With regard to physical data management, the question remains whether components that are part of the same larger structure are also stored close together on the disk. Various commercial OODBMS offer ways to achieve such a spatial clustering, at least to a some degree.

Several classes in the example given above (Figs. 3.32–3.35) use the concept of structural object-orientation. The class *City*, for example, has an attribute *Districts*, which is an enumeration of its districts and therefore set-valued. Cities, districts, and roads all have an attribute *Shape*. *Shape* is of type *SpatialSet* in the case of *City*. For *District* and *Road*, this type was overwritten by the types *PolygonSet* and *LineSet*, respectively.

Figure 3.36 illustrates these concepts. Pointers to members of other classes are represented by the syntax *Class<ID>*. In a real implementation they would be represented by the corresponding OID. Set-valued attributes are indicated by curly brackets. For brevity, we only show some of the spatial object classes. Polygons are represented by vertex lists, and vertices are in turn represented by their coordinates $X$ and $Y$. These components are integers and therefore atomic.

### 3.4.4 Behavioral Object-Orientation

A data model is called *behaviorally object-oriented* if it supports user-defined types, and operators (in this context usually called *methods*) that are applied to these types. Behavioral object-orientation is a direct outcome of the work on integrating abstract data types into databases (cf. Sect. 3.2.3). After specifying a type and its associated operators, users can treat the new type just like a system-defined type. Operators are defined by an interface or *signature* (name, input parameters, output parameters) and a program to compute it. The OODBMS guarantees that operators cannot be applied to inappropriate objects. To apply an operator, one only needs to know its interface; no information is required about its implementation.

Typically, users have little interest in the implementation details of the data types and operators. On the contrary, they usually prefer an encapsulation approach, where abstract operators are the only way to manipulate or communicate with the objects. Encapsulation allows one to specify whether an attribute or method is visible to the user. Moreover, one can define different ways these attributes and methods may be accessed. In $O_2$, for example, an attribute may be specified as *private* (by default), *public*, *read*, or *write*.

For a simple example, let us consider the class *Spa*. The following method *modify* changes the *Tax* attribute by a specified amount:

```
method modify(Difference: integer) in class Spa
       body (*self.Tax += Difference);
```

The keyword *method* starts the definition of *modify*, which has an input parameter *Difference* of type *integer* and no output parameter. The method is to be applied to spas only. After the definition of the signature, the keyword *body* denotes the beginning of the corresponding program. The tax corresponding to the object in question (*self*) is increased or decreased by *Difference*.

City

| Name | Shape | Population | Mayor | Districts |
|------|-------|-----------|-------|-----------|
| San Francisco | SpatialSet<7> | 590 000 | Brown | {Russian Hill, Nob Hill, SOMA, ...} |
| Berlin | SpatialSet<9> | 3 800 000 | Diepgen | {Charlottenburg, Kreuzberg, ...} |

District

| Name | Population | Shape |
|------|-----------|-------|
| Russian Hill | 10 000 | PolygonSet<17> |
| Nob Hill | 12 000 | PolygonSet<19> |
| SOMA | 20 000 | PolygonSet<21> |
| ... | ... | ... |

Road

| Name | Shape |
|------|-------|
| Hyde Street | LineSet<25> |
| ... | ... |

PolygonSet

| ID | set(Polygon) |
|----|--------------|
| 17 | {Polygon<25>, Polygon<27>, ...} |
| ... | ... |

Polygon

| ID | list(Point) |
|----|-------------|
| 25 | {Point<34>, Point<37>, ...} |
| ... | ... |

Point

| ID | X | Y |
|----|---|---|
| 34 | 7 | 15 |
| 37 | 9 | 11 |
| ... | ... | ... |

**Fig. 3.36.** Structural object-orientation

The next example refers to the class *City*. The following methods serve to compute the shape of a city from the corresponding district information. It is assumed that an operator *union* has already been defined that computes the union of polygons. Here it is used to compute the shape of the union of several neighboring districts from their individual shapes. The implementation of this operator works directly on the internal representations of the object's shapes.

```
method comp_pop(): integer in class City
      body (d: District
              p: integer
              for each d in *self.Districts (
                    p = p+d.Population)
              return(p));
```

```
method comp_shape(): PolygonSet in class City
      body (return select union(District.Shape)
                  from District in *self.Districts);
```

The code below describes a more complex example that involves the two classes *Species* and *BiotopeInhabitant*. Species such as frogs, waders, and marsh plants are found in particular biospheres, for example in or beside rivers or in marshes. To determine the locations of the species found in a given biotope, we define a method *to_be_found* in *BiotopeInhabitant* that returns the spatial representation of their occurrence. Remember that each element of the class *Biotope* has an associated spatial representation (the *Shape* attribute). The method *to_be_found* computes the spatial locations for a given *BiotopeInhabitant* by computing all intersections between its corresponding bio*spheres'* shapes and its bio*tope's* shape. The method *cover* is a spatial operator that forms the union of all partial results and turns it into one object of type *SpatialSet*.

```
class Species
      public type tuple(
                    Name: string,
                    Biosphere: set(EnvObject))
      end;
```

```
class BiotopeInhabitant inherit Species
      public type tuple(
                    Home: Biotope)
      method to_be_found: SpatialSet
      end;
```

```
method body to_be_found: SpatialSet in class BiotopeInhabitant
      {
      o2 EnvObject env;
      o2 set(SpatialSet) tempset;
      for (env in *self->Biosphere)
         tempset
         += set(env->Shape->inter(*self->Home->Shape));
      return cover(tempset);
      };
```

In addition to application modeling, behavioral object-orientation is also an efficient tool for integrating spatial operators into a database system. The following example serves to declare several methods associated with the class *Line*. The method *length* computes the length of the object it is called on (*self*), and *inter(l)* computes the intersection of *self* and the line *l*. *length* refers to the list operators *head* and *succ* that compute the head of a list and the successor of a given element, respectively.

```
method public intersection(l: Line): Line in class Line;

method public length: real in class Line;

method body length: real in class Line {
        r = *self.head;
        while r.succ do {
                result += sqrt (sqr(r.succ->Y - r->Y)
                              + sqr(r.succ->X - r->X));
                r = r.succ
        };
```

The integration of user-defined types and operators immediately leads to more powerful query languages. For example, if the user defined a class *Line* and some associated methods such as *inter* or *length*, one can now use these constructs in $O_2$SQL, the ad hoc query language of $O_2$ as follows:

```
select r
from   r in Line
where  r->length > 5.0

select r->length
from   r in Line
```

The first query returns a set of rectangles, each having an area greater than five. The second query returns a set of real numbers, each corresponding to the area of a rectangle. Similarly, methods can be used in an embedded mode or directly in $O_2$'s C-based programming language $O_2$C:

```
run body {
        o2 set(Line) result;
        o2 real minarea;
        minarea = 5.0;
        o2query(result, "select r from r in Line
                        where r->area > $1", minarea);
        display (result);
        };
```

```
run body {
        o2 Line r;
        o2 set(real) result;
        for (r in Line)
            result += set(r->length);
        };
```

### 3.4.5 Overloading

*Overloading* means that one may attach different implementations to a single function name or type specification. Depending on the argument types, the system then decides at compile time or run time (*late binding*) which implementation to use. In the following example, the method *init* is declared as writing values to an object of class *Point*. Note that for each class there is a method *init* supplied by the system, which is called when creating a new object with the *new* operator. By overloading *init*, users can implement their own initialization routine, e.g., in order to include an external transformation procedure.

```
method write init(X: real, Y: real) in class Point;
```

```
method body init(X: real, Y: real) in class Point {
        *self->X = X;
        *self->Y = Y;
        };
```

In the next example, we define a method *modify* on class *UnivTown* to update the number of students.

```
method modify (Difference: integer) in class UnivTown
        body (*self.NoStudents += Difference);
```

Although there already exists a method with the same name in class *Spa*, there is not necessarily a conflict. Because of encapsulation, the objects of *Spa* do not know the methods defined in *UnivTown*. Only for objects that belong to both classes, such as *Stuttgart*, one has to provide a mechanism to decide which method should be used.

A similar example can be given for the method *comp_shape*, which already exists for the class *SpatialSet*:

```
method comp_shape(): PolygonSet in class District
        body (return *self.Shape);
```

If *comp_shape* is now applied to a district rather than a whole city, it just returns the value of the corresponding *Shape* attribute.

If one has multiple representations for certain classes of objects, one can use overloading to define specific functions and access methods for each representation. If one needs to compute the length of a line, for example, one may do that using either its vector or its raster representation. In the raster format, this turns out to be a rather cumbersome operation, whereas the length of a vector line is very easy to compute. On the other hand, the intersection of two areas represented as vector lists is much more difficult to compute than the intersection of their raster representations. It becomes an interesting optimization problem to decide which representations and which access methods to use in a given situation [BG92].

### 3.4.6  GODOT: An Object-Oriented GIS

The main idea behind the GODOT project [GR93] was to develop the prototype of a geographic information system on top of a commercial object-oriented database system by adding appropriate classes and methods. The project was conducted between 1992 and 1996 at FAW Ulm. It was funded by the Environmental Ministry of the State of Baden-Württemberg and Siemens Nixdorf Informationssysteme AG.

With its strictly object-oriented approach, GODOT differed from most other GIS developments at the time in various respects:

- GODOT's object-oriented data model allows the representation of complex geographic and environmental data objects.
- The GODOT data model is extensible by user-defined classes and methods.
- GODOT's underlying OODBMS allows both spatial and non-spatial data to be stored in an integrated manner.
- GODOT is based on a commercial OODBMS and therefore automatically participates in new commercial developments. This may concern features such as query language standards, graphical tools, transaction management, and distributed processing.

Earlier related work was mostly based on database research prototypes. The PROBE project discusses spatial data modeling and query processing aspects in an image database application [OM88]. Scholl and Voisard implement GIS functionalities on top of a pre-commercial version of the OODBMS $O_2$ [SV92]. Schek et al. use their extensible database system DASDBS as a foundation for an advanced GIS prototype [SPSW90]. Oosterom and Vijlbrief present a GIS implementation based on the extensible database system POSTGRES [OV91]. The Sequoia 2000 project constitutes a large-scale effort to use POSTGRES for the management of geographic and environmental data [Sto93, Fre94, FD97]. Shekhar et al. [SCG+97] propose a GIS data model that unifies the object-oriented approach with the classical layer model (cf. Sect. 3.1).

Both POSTGRES and DASDBS can be regarded as object-relational database systems: they integrate object-oriented features with the traditional relational data model, thus avoiding a major paradigm change and associated legacy problems. This approach has been widely recognized as a promising option to bring object-oriented techniques to the market at large.

In the subsequent presentation of GODOT, we follow the article by Günther and Riekert [GR93]. We begin with a discussion of the system architecture, followed by a description of the data model.

**System Architecture.** The GODOT system has a four-layer architecture (Fig. 3.37), consisting of:

1. the *commercial OODBMS ObjectStore* [LLOW91, Obj98]
2. an extensible *kernel* with classes and methods for the representation and management of complex spatial and non-spatial data objects
3. a collection of *base components* for query processing, graphics, database administration, and data exchange
4. several *user interface modules*, including a C/C++ program interface, a UNIX command interface, and a graphical user interface based on X Windows and OSF/Motif.

In the following we discuss these four layers in turn.

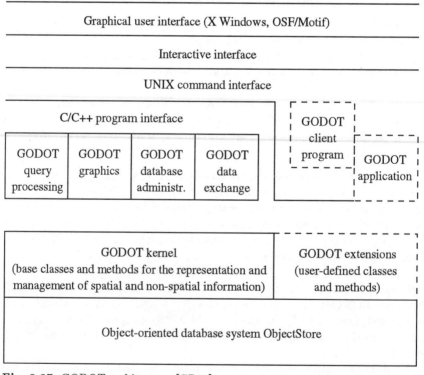

**Fig. 3.37.** GODOT architecture [GR93]

GODOT is implemented directly on top of the *commercial OODBMS ObjectStore* [LLOW91, Obj98]. ObjectStore is a fully object-oriented database system in the sense of the Object-Oriented Database Manifesto [ABD+89]. This covers in particular the basic database functionalities, including transaction management and indexing. Some of these functionalities have been adapted and extended for more efficient management of spatial data.

The GODOT *kernel* contains the definition of special classes and methods for the management of spatial and environmental information. In particular, the implementation of the GODOT data model is located here (see Sect. 3.4.6). The data model can be extended by classes and methods to customize the system to particular applications. Also located in the kernel are the spatial clustering and indexing components. Like all the other GODOT components, the kernel was implemented in C++.

GODOT's *query component* contains several GIS-specific language elements to enhance the query language of ObjectStore. The enhancements are partly implemented using a feature of ObjectStore that allows to call user-defined methods from within a query. The query component interacts directly with the interface modules described below.

The system's *graphics component* serves to visualize the result of user queries, to specify parts of a query by pointing to certain objects on the screen, and to update geographic information by manipulating the corresponding graphic objects interactively.

The *database administrator component* provides the usual features for database schema manipulation, user administration, and system support. While some of ObjectStore's functionalities could be used directly, others had to be adapted to the specific requirements of spatial data management.

To support the exchange of spatial data encoded in different formats, GODOT supports the integration of *data interfaces* as base components. They can be activated through any of the interface modules described below. GODOT also has its own external data format, which is a subset of C++, encoded in ASCII. This external data format can be read easily by other systems. The execution of the code leads directly to the generation of the corresponding object classes and instances in the given database.

One of GODOT's core functionalities is to be a GIS data server for a diverse and distributed collection of applications. For this purpose, GODOT offers a variety of client-server style *user interfaces*. An *interactive interface* under X Windows gives high-level graphical access to GODOT, especially for the occasional or non-expert user. Another kind of access is provided by the *UNIX command interface*, where GODOT queries can be formulated by means of specialized commands that form an extension of the UNIX shell. Command procedures can then be implemented as shell scripts. Finally, a *C/C++ program interface* makes the GODOT modules available as a program library. This interface is typically used for more complex GODOT applications; it also allows remote access via the Remote Procedure Call (RPC) protocol. The C/C++ program interface has been implemented in compliance with the object request broker recommendations of the Object Management Group [Obj95a, Obj95b].

**Data Model.** The GODOT data model distinguishes between three kinds of objects: thematic objects, geometric objects, and graphic objects. The relationships between these kinds of objects are shown on a simple example in Fig. 3.38. The directed links denote *part-of* relationships, the other links represent different semantics (usually application-dependent). We again use strings of type *Class<ID>* as object identifiers. There are three thematic objects in the example: two objects *City<Ulm>* and *City<Neu-Ulm>* to represent the twin cities Ulm and Neu-Ulm, and an object *CoordCommittee<31>* to represent the coordination committee of the two cities. The thematic objects representing the two cities are connected to geometric objects to represent their shapes and to graphic objects for their cartographic representation. We now discuss the three types of GODOT objects in more detail.

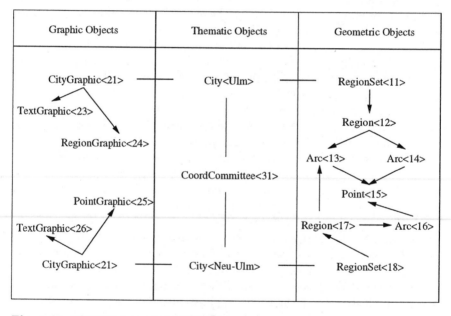

| Graphic Objects | Thematic Objects | Geometric Objects |
| --- | --- | --- |

**Fig. 3.38.** GODOT data model [GR93]

*Thematic objects* roughly correspond to the *environmental data objects* introduced in Chap. 1 of this book. Thematic objects are used to represent real-world objects. They may be simple or complex, i.e., composed of several other thematic objects. The coordination committee and the cities in Fig. 3.37 are typical examples of thematic objects.

Thematic objects may have different kinds of attributes to represent geographic and environmental features:

1. Attributes of an elementary type (e.g., text strings or real numbers)
2. Attributes of a complex type (e.g., embedded classes in C++)
3. Attributes of a referential type (e.g., pointers to other thematic objects).

The most important subset of thematic objects is formed by the *geographic objects* or *geo-objects*, which are characterized by an attribute of some spatial data type. A geo-object therefore has a location and a spatial extension. If a geo-object is complex, a number of integrity constraints need to be enforced. For example, the geometry of a complex geo-object has to be the union of its component geometries.

Associated with each geo-object is a *geometric object*, which may be elementary or complex. Elementary geometric objects belong to the class *Spatial* or one of its three subclasses *Region, Arc,* and *Point.* Between these classes there exist the usual spatial relationships. A region has several bounding arcs, and an arc may in turn belong to any number (including zero) of regions. Similarly, an arc has two endpoints, and any point may be the endpoint of any number of arcs. Note that in extension of the data model presented in Sect. 3.2.1, an *arc* may be curved, whereas a *line* was defined to be piecewise linear. As a consequence, a *region* may have a curved boundary, which was of course not possible for a *polygon.*

Complex geometric objects that are composed of only regions belong to the class *RegionSet.* The classes *ArcSet* and *PointSet* are defined analogously. If a complex object is heterogenous in the sense that it contains components from different classes, it belongs to the class *SpatialSet,* which is the superclass of *RegionSet, ArcSet,* and *PointSet.*

This design guarantees that the result of a boolean set operation on two geometric objects can always be represented by exactly one geometric object. It also enables us to represent the geometries of any geo-object, however oddly shaped, with just one geometric object. Consider, for example, a river whose width varies widely, such that its geometry has to be represented by both arcs and regions. In GODOT, the geometry of this river would be modeled by an instance of the class *SpatialSet.* Another example is a country whose area consists of several disconnected regions (e.g., the United States), which would be represented by an instance of *RegionSet.*

In the example shown in Fig. 3.38, each city's shape is modeled by a geometric object of type *RegionSet.* In the example, both sets have only one element. The two regions 12 and 17 share the line 13, one of whose endpoints is the point 15. This point is also the endpoint of line 14 (which belongs to region 12) and line 16 (to region 17). Figure 3.39 shows an example topology that follows these specifications.

*Graphic objects* are used for the (interactive or printed) visualization and the interactive update of thematic objects, in particular of geo-objects. The looks of a graphic object are defined in detail by its attributes, such as color, line width, or text font. Possible graphic representations include business graphics, tables, videos, raster images, or GIS-typical vector graphics. A the-

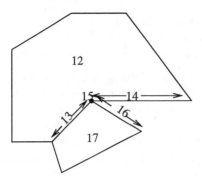

**Fig. 3.39.** Geometric objects in GODOT

matic object can be linked to several graphic representations (e.g., for multiple scale display).

An important subcategory of graphic objects is formed by the *cartographic objects*, which are used for the graphic display of geo-objects. With this design, GODOT maintains a clear separation between geographic and cartographic information. Cartographic objects contain methods that determine how the properties of a given geo-object are represented in terms of the graphics available. In particular, questions of scale and cartographic generalization are handled at this level and *not* at the level of thematic objects. This clear separation is similar to Egenhofer's design of Spatial SQL [Ege91, Ege94], where all graphical aspects are handled in a separate logical environment (GPL, cf. Sect. 3.2.4).

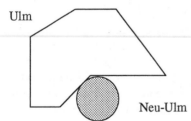

**Fig. 3.40.** Cartographic objects in GODOT

## 3.5 Summary

In this chapter, we gave a comprehensive overview of storage techniques in geographic information systems and spatial databases. We began with an overview of data storage in GIS (Sect. 3.1), pointing out some of the notorious problem areas. Classical GIS, in which the data is administered in simple file systems, have major difficulties in efficiently managing the large amounts of data that are typical especially for environmental applications. Furthermore, increasing user requirements such as structured object modeling, concurrency, and recovery are difficult to realize in traditional GIS.

Classical database management systems, on the other hand, also seem unable to support complex geographic and environmental applications in an efficient manner. Due to recent research results, however, this is about to change in the near future. Spatial database research has developed powerful techniques to handle complex geometric data types and operators. Work on object-oriented and object-relational database systems provides the means to integrate spatial data management techniques into a commercial setting. This flexibility is a clear advantage compared to the fixed set of data types and operators that is typical for classical commercial DBMS. It can be used in a variety of ways to enhance the functionality and efficiency of an environmental information system.

Section 3.2 described the efforts of database researchers to integrate spatial data types and operators into a classical database framework. Section 3.3 focused on a specific problem area of spatial databases: the development and evaluation of efficient multidimensional access methods. Section 3.4 was devoted to object-oriented approaches to spatial data management. We also presented GODOT, a research prototype for a GIS implemented on top of a commercial object-oriented DBMS.

We conclude that modern database technology is essential for the efficient handling of geographic and environmental data. For the necessary integration of GIS and modern database technology, there are essentially four options:

1. *Extension of an existing GIS by database functionalities.* Earlier versions of Siemens Nixdorf's GIS SICAD are a typical example of this approach. SICAD's proprietary data management system GDB offers advanced database functionalities and serves as integrated storage for both spatial and the non-spatial data.

2. *Coupling of a GIS with a commercial DBMS.* Such a coupling has been common for some time for storing the non-spatial data in a commercial relational database. ESRI's ARC/INFO, for example, has been offering this as an alternative to its proprietary INFO component. Since the mid-1990s, many vendors started to store spatial data in a relational DBMS as well. In SICAD/open, for example, GDB has been replaced by a solution called GDB-X [Lad97] that relies on commercial relational database systems. As in previous SICAD versions, both spatial and non-spatial data are stored in the same database. For the spatial data, however, the relations just serve as containers that manage the geometries as unstructured long fields. ESRI provides a similar solution with its Spatial Database Engine (SDE) [ESR98, ESR97c].

3. *Extension of an existing DBMS by geometric and geographic functionalities.* For this approach the use of an OODBMS seems to be an interesting option. Early prototypes were based on POSTGRES [OV91], O$_2$ [SV92], or ObjectStore [GR93]. The object-relational database company Illustra,

which was later bought by Informix, is a successful commercial example. Domain-specific extensions are packaged in specific modules called *DataBlades* [Gaf96, Inf97].

4. *Open toolbox approaches* that see a GIS just as a collection of specialized services in an electronic marketplace [VS94, VS98]. While there is currently no commercial system that strictly follows this architecture, many vendors are starting to integrate similar ideas into their products. GeoServe of Siemens Nixdorf [Lad97] and ESRI's OpenGIS proposals [ESR97a] represent first steps in this direction.

# 4. Data Analysis and Decision Support

In this chapter, we discuss how the stored environmental data can be prepared for decision support purposes. This usually involves some more aggregation and detailed analysis of the available data. Compared to the techniques presented in previous chapters, decision support is more target-oriented in the sense that it takes the particular requirements of a given decision-making task into account. As a result, the selection of data sets and analysis techniques is more application-specific than in the case of, say, raw data processing. Related work goes back into the late 1980s [GW89a]. The topic is increasingly attracting attention among both environmental and management scientists.

We begin with a description of environmental monitoring, which is a particularly important application of data analysis and decision support in the environmental sciences. Section 4.2 gives an overview of simulation models for environmental applications. Section 4.3 lists some of the data analysis tools offered by commercial geographic information systems. Section 4.4 discusses online databases and the impact of the World Wide Web on environmental information management. Section 4.5 gives an overview of environmental information systems in the enterprise. We conclude with a case study: Section 4.6 presents UIS, the environmental information system of the Southern German state of Baden-Württemberg.

## 4.1 Environmental Monitoring

The collection and monitoring of environmental data is an essential component of any environmental management and protection strategy. Environmental monitoring consists of a continuous evaluation of the incoming data streams. Fritz [Fri92] and Günther et al. [GRR95] give an overview of related international activities. The purpose is to recognize unusual developments early on to avoid serious damage. Unusual developments may be event-oriented, such as a sudden increase in the concentration of a dangerous substance, or gradual, such as the slow change of a river's topology. Both kinds of developments are potentially dangerous. Once such a development has been spotted, one has to obtain more detailed information on the potential sources and, possibly, initiate countermeasures.

Most governmental agencies employ a multi-level approach for environmental monitoring. Management of the measuring stations and control of the primary data capture is usually performed by local government agencies, i.e., agencies at the city and county level. Those agencies use the data for tactical tasks, such as short-term resource distribution, quality assessment, pollution detection, and treatment monitoring. Aggregate data is forwarded to state, national, and possibly international agencies for middle- and long-term strategic planning, including policy assessment and legislation.

With regard to the data flow sketched in Chap. 1, this process forms a natural continuation of the data capture and storage phases. It corresponds to a further semantic aggregation, as symbolized by the pyramid in Fig. 1.2. At the technical level, this process can be supported by a modular system architecture. Pham and Wittig [PW95], for example, propose a system design for river quality monitoring that is based on four functional entities (Fig. 4.1):

1. Sensor data acquisition
2. Validation procedures
3. Situation description
4. Situation assessment.

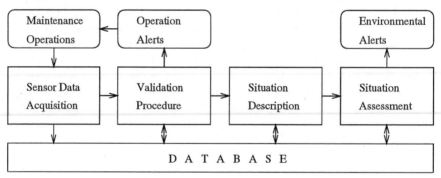

**Fig. 4.1.** Functional modules of a river quality monitoring system [PW95]

The first two of those four entitities are part of what we termed *data capture*. In the third stage, Pham and Wittig propose to go beyond measurement values in order to obtain a description of the overall situation. This step involves the computation of derived values based on multiple measurements, comparisons with historical data, and similar data fusion operations.

At some point, human judgement will be introduced with the objective of obtaining a coherent assessment of the environmental situation and ongoing developments. If necessary, an environmental alert will be triggered. This last step, termed *situation assessment*, is hard to separate from the situation description stage, as human judgement usually enters the analysis in a gradual manner. In most applications, human judgement is required before a major alert is triggered. This design reflects the current state of related software tools. While useful for the low-level acquisition and aggregation procedures, there rarely exists a situation where the high-level analysis and decision-making can be left to a computer. Given the complexity and inherent uncertainty of environmental data, as well as the political dimension of environmental decision-making, human judgement will remain indispensable for the foreseeable future.

This whole sequence of steps is supported by a comprehensive database or, more typically, a federation of several location- and task-specific databases. These databases contain all the measurement data, procedural data, and aggregate data that is used and generated as part of the process.

It is the last two functional modules that we want to emphasize in this chapter: situation description and assessment. Both tasks require some higher-level synthesis of the available data and, frequently, an evaluation by human experts. During those stages it is sometimes difficult to separate the data analysis from the decision making itself because the analysis is performed with a particular agenda in mind. If the various steps of the analysis are distributed among a hierarchically organized group of people, such as a government agency, this phenomenon becomes very obvious. Everybody in the hierarchy may have a certain bias on what is important, which may have an impact on the kind of information that is sent higher up. This is not necessarily a bad thing. After all, that is what expert counsel is all about: the selection and analysis of what the expert deems significant. It is important, however, that decision makers are aware of the various sources of error and bias that the information has been exposed to before being made available to them.

## 4.2 Simulation Models

### 4.2.1 Background

Models and simulation have long been an important analysis tool in a variety of disciplines. A *model* is an abstract description of a real-life phenomenon. The abstraction usually involves some simplification and results in a formal representation. To obtain insights about the underlying real-life phenomenon, one often performs experiments with the model, hoping that the results of such a *simulation* somehow correspond to the behavior of the system in real life. Many models have been developed specifically for performing simulations on them; such models are termed *simulation models*.

Simulation models have traditionally been one of the most important and demanding computer applications. Especially in the early age of the computer, these applications have been a driving force behind the development of ever more powerful hardware. Even today, simulations are one of the most common applications for supercomputers. To support such computations, there exists a variety of specialized programming languages (such as SIM-ULA) and commercial simulation software environments (such as Simulink [Sci97b]). Figure 4.2 shows the various connections between modeling, simulation, software implementation, and validation.

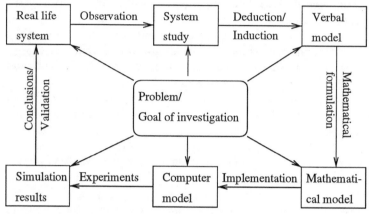

**Fig. 4.2.** Modeling and simulation [GHP95]

Arguably the most notorious applications of simulation models are population dynamics and weather forecasting. Simulation models for population dynamics go back to the 1920s, when Lotka and Volterra formulated their model to simulate the interdependencies between two adverse populations (see, for example, [Ric85]). Typical parameters entering the computation are food supply, or birth and death rates. Meteorological simulation models, usually based on the Navier–Stokes equations on the preservation of mass and impulse, are of more recent origin.

### 4.2.2 Environmental Applications

From those foundations, environmental information science has developed a great variety of models for different applications. Sydow [Syd96a] lists, among others, the analysis and prevention of acid rain, the evaluation of environmental consequences of a nuclear strike ("nuclear winter"), and the analysis of carbon dioxide and ozone concentration in the stratosphere. Some of these applications help to explain general principles underlying our ecosystem. Others serve specifically to forecast the effect of certain intended actions. Especially the latter kind of models – also called *scenarios* – are of obvious importance to planners and decision makers.

The field of environmental modeling has a long history, and it is beyond the scope of this book to cover this field in great depth. Good starting points for further reference include the textbook by Bossel [Bos94], two collections of articles edited by Goodchild et al. [GPS93, GSP96], and the survey articles by Fedra [Fed93, Fed94] and Hilty et al. [HPRR95]. Readers of German may also refer to a survey article by Grützner et al. [GHP95] and to a recent collection edited by Grützner [Grü97],

There also exists a comprehensive online listing of simulation models in environmental applications: the Register of Ecological Models (REM), maintained by a research group at the University of Kassel (Germany) [BV95, BV96]. REM is available at http://dino.wiz.uni-kassel.de/ecobas.html. It allows users to search for models by name or by content. Content-oriented search can be structured according to the medium the model is concerned with (*air, terrestrial, freshwater,* or *marine*), and its main subject (*bio-/ geochemical, population dynamics, hydrology, (eco-)toxicology, meteorology, agriculture,* or *forestry*). Figures 4.3 and 4.4 illustrate the use of the system.

In the remainder of this section, we briefly describe the model types that are most important in environmental applications and the problems specific to this application domain. Following Grützner et al. [GHP95], we distinguish between four classes of simulation models that are particularly relevant for environmental applications: transport models, resource utilization models, process models, and ecosystem models. We discuss the four model types in turn.

**Transport Models.** Transport models simulate the migration of substances in the air, the water, or the ground. The most sophisticated transport models have been developed for air transport (see, for example, [Bar95] or [Syd96b]). One distinguishes between *Eulerian models* that rely on grid-based computations and *Lagrangian models* that consider single particles. Transport models start from assumptions about the emissions and conditions of transport. From there they make predictions about the immissions at particular locations. Transport models serve to approximate immissions in places where no measuring stations can be installed, and to forecast immissions resulting from some intended action (e.g., the construction of a factory) or an unintended emergency. Many models can also be used in an inverse manner. If one observes some immissions that are beyond the legal limit or otherwise unusual, one can thus find out the potential source of those immissions.

**Resource Utilization Models.** For some classes of resources, there are models that go beyond the transportation aspect to take a broader variety of related phenomena into account. For water in particular, there exists a large body of related work. Quantitative groundwater models help to forecast possible changes in the amount and flow of the groundwater following construction and similar actions. Other models serve to ensure water quality in a given lake and river system [ATK97]. Resource modeling requires the consideration of a large variety of external factors, including the local geology, weather conditions, and human interaction.

**Process Models.** Process models serve to simulate and optimize technical processes [GM88]. Measurements taken on existing real-life processes serve

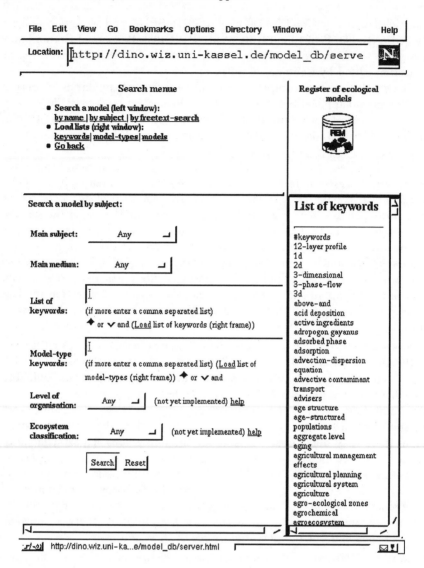

**Fig. 4.3.** Register of Ecological Models (REM)

File    Edit    View    Go    Bookmarks    Options    Directory    Window    Help

Location: http://dino.wiz.uni-kassel.de/model_db/mdb/przm2

## I. General Model Information

Name: Pesticide Root Zone Model

Acronym: PRZM2

Main medium: terrestrial
Main subject: hydrology, biogeochemical
Organisation level: ecosystem
Type of model: 1 D, partial differential equations (PDE) – finite elements
Keywords: pesticide, root zone, solute transport, agriculture, unsaturated soil, richards equation,
convection–dispersion equation, finite element numerics, finite difference numerics, Monte–Carlo–simulation,
vapour phase transport, biodegradation, MOC

Contact:

*Center for Exposure Assessment Modeling (CEAM) (EPA)*

Authors:

Carsel, R.F., C.N. Smith, L.A. Mulkey, J.D. Dean

Abstract:

PRZM (Pesticide Root Zone Model) is a finite–difference model simulates the vertical one–dimensional
movement of pesticides in the unsaturated zone within and below the root zone. The model consists of
hydrologic (flow) and chemical transport components to simulate runoff, erosion, plant uptake, leaching, decay,
foliar washoff, and volatilization. Pesticide transport and fate processes include advection, dispersion,
molecular diffusion, and soil sorption. The model includes soil temperature effects, volatilization and vapor
phase transport in soils, irrigation simulation and a method of characteristics algorithm to eliminate numerical
dispersion. Predictions can be made for daily, monthly or annual output. PRZM allows the user to perform
dynamic simulations considering pulse loads, predicting peak events, and estimating time–varying emission or
concentration profiles in layered soils. PRZM2 links two subordinate models in order to predict pesticide fate
and transport through the crop root zone, and the unsaturated zone: PRZM and VADOFT. PRZM, VADOFT
and SAFTMOD are part of RUSTIC. RUSTIC links these models in order to predict the fate and transport of
chemicals to drinking water wells. The codes are linked together with the aid of a flexible execution supervisor
(software user interface) that allows the user to build models that fit site–specific situations. This release of
PRZM incorporates several features in addition to those simulated in the original PRZM code: specifically, soil
temperature simulation, volatilization and vapor phase transport in soils, irrigation simulation, microbial
transformation, and a method of characteristics (MOC) algorithm to eliminate numerical dispersion. PRZM is
now capable of simulating fate and transport of the parent compound and up to two daughter species. VADOFT
is a one–dimensional finite–element code which solves the Richard's equation for flow in the unsaturated zone.
The user may make use of the constitutive relationships between pressure, water content, and hydraulic
conductivity to solve the flow equations. VADOFT may also simulate the fate and transport of two parent and

file://ftp.epa.gov/epa_ceam/wwwhtml/ceamhome.htm

**Fig. 4.4.** REM entry of the Pesticide Root Zone Model PRZM2

as input to such models, which then compute various scenarios with different parameter sets. Based on the simulation results, one obtains an optimized set of parameters, which can then be validated and used as input to the real-life process. Typical applications include the simulation of combustion engines, the modeling of chemical plants, and the optimization of power plants.

Besides improving the overall economic efficiency of a process, such an optimization may have positive consequences with regard to the environment. On the one hand, it may decrease the input required to produce a specified output. This may result in saving natural resources; increasing the fuel efficiency of an engine is a typical example. On the other hand, it may lead to a decrease in emissions, i.e., in unintended and unusable output. Finally, optimization may improve overall safety standards, thereby decreasing the chance of accidents and the resulting environmental damage.

**Ecosystem Models.** The objective of ecosystem models is to describe and quantify the effect of substances on living organisms. Eventually one attempts to determine the stability and elasticity of a given ecosystem, i.e., its ability to absorb external substances without going out of kilter.

Following [GHP95] we distinguish between *compartment-oriented* approaches and *individual-oriented* approaches. The idea of a *compartment* is similar to the concept of a class in object-oriented software design: similar objects are grouped into sets that can then be viewed at a more abstract level. For example, the organisms belonging to a given species or population can be grouped into a compartment and then be described as one abstract entity. The Lotka–Volterra models mentioned above are based on this principle: their populations are compartments in the given sense. Another common application area of compartment-oriented approaches is forest modeling.

If this kind of abstraction is undesirable, one may want to use an ecosystem model that is more oriented towards the *individual organism* and its interactions with other individuals. Each individual with its possible actions between birth and death is then modeled as a separate entity. Such approaches – sometimes termed *artificial life* – have been successful in the modeling of population dynamics and evolution, survival strategies, and social optimization strategies.

### 4.2.3 Problem Areas

A key problem in environmental simulation models concerns the match between the data available and the data required by a given model. Environmental measurements are not standardized; they differ both in *what* they capture and *how* they capture it. As for the latter aspect, there are different measuring techniques but also more mundane differences that could easily be harmonized, such as scale and frequency of measurements. The United Nations Environmental Program (UNEP) has established a task force to work on this problem, called UNEP-HEM (Harmonization of Environmental Measurements). Keune et al. [KMB91] give an overview of the group's mission and concrete plans. While their discussion of data availability problems is somewhat obsolete by now, due to the rise of the Internet, their description of syntactic and semantic heterogeneities is as current now as it was back in 1991.

Another notorious problem area concerns the models' user interfaces and visualization features [Den93, GHP95, DMH95]. Many important models have been developed incrementally over several decades. They are typically programmed in Fortran and run in a mainframe-based computing environment. In addition, they are often badly documented – in summary, a typical legacy software dilemma. Making such models more user-friendly, easier to maintain, and compatible with open client/server architectures provides a long-term challenge for the modeling community.

The World Wide Web and related tools (such as the Common Gateway Interface [McC94]) provide additional incentives to do so because they greatly facilitate the distribution of software tools. Models coupled with state-of-the-art user interfaces and visualization tools could be packaged as interactive services and made available to the community via the Internet, either as a free service or as a commercial product. For the older generation of Fortran- and mainframe-based simulation software, this will be difficult to achieve. Reimplementations will often be necessary to enter this new era of "Internet marketplaces" [Abe97].

### 4.2.4 DYMOS/DYNEMO: A Model for Smog Analysis

We conclude with a short case study: the DYMOS/DYNEMO simulation system developed by Sydow et al. [Syd94b, SLMS97]. The main objective of DYMOS/DYNEMO is to give decision support for environmental planners and administrators. The system also provides input to public smog forecasts.

DYMOS (DYnamic MOdels for Smog analysis) is a complex dispersion model to analyze the flow of hazardous substances in the lower atmosphere [Syd94a]. The model is mainly geared towards smog (especially in connection with high ozone concentrations) and antigene pollutants (which affect the human immune system). DYMOS is composed of an atmospheric model, a Eulerian transport model, and a tropospheric gas phase chemistry model called CBM-IV. CBM-IV serves to simulate certain aerial chemistry phenomena; it was originally developed by the U.S. Environmental Protection Agency [GWK88]. In addition, DYMOS is integrated with a traffic flow model called DYNEMO to simulate the impact of traffic flow and car emissions in the context of smog analysis.

Figure 4.5 shows the resulting system architecture. Relevant system components include a graphical user interface and a database. From the interface, the user can start simulations, review state variables, and visualize simulation results in two and three dimensions. The project team is currently working on a more sophisticated visualization tool that allows users to review the simulation results as if one were flying through the area of interest.

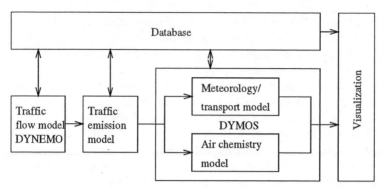

**Fig. 4.5.** DYMOS/DYNEMO simulation system for smog analysis [SLMS97]

Simulation models like DYMOS often face considerable data acquisition problems. In DYMOS the geographic input data is obtained from digital maps and satellite imagery. It usually requires manual preprocessing to obtain exactly those formats than can be understood by the model. Another problem concerns the dynamization of aggregate data on emissions and traffic flow. In the case of emissions, for example, one may have only annual statistics. Based on such aggregates and certain assumptions on the underlying distributions, one then estimates the corresponding measurements for much shorter time periods; DYMOS usually requires data at a granularity of days or even hours. A similar problem exists in the case of traffic input data. Here the model requires data that is classified on the basis of road segments, day of the week, and time of day. This kind of accuracy is usually not available. Rather than just using simple estimates, however, DYMOS relies on more sophisticated results from the connected traffic flow model DYNEMO to simulate the required accuracy.

In other words, one uses one simulation model (DYNEMO) to generate the input data for another simulation model (DYMOS). This approach is not uncommon, as it is often the only way to obtain interesting results despite coarse or missing input data. Once again, however, it is important to keep track of the uncertainties introduced by this technique, and to make these uncertainties apparent to decision makers who base their decisions on the model output.

## 4.3 Data Analysis in GIS

While data analysis has always been an important part of GIS, the breadth and depth of related work has increased considerably since the early 1990s. By extending their functional spectrum beyond the traditional domains of data capture, storage, and visualization, GIS are gradually moving into the mainstream of computing. Rather than providing support just for the geosciences, GIS vendors are trying to position their products as spatial data management components that should be a part of just about any information system architecture – simply because just about any information has a spatial aspect. Interfaces to business software such as SAP's R/3 [SAP98] and the development of *spatial decision support systems* [Den91, AYA+92] are among the most visible signs of this trend.

In order to achieve these ambitious goals, GIS vendors have to provide data analysis capabilities that go far beyond simple map overlays (cf. Sect. 3.1). Openshaw [Ope91] gives a systematic overview of possible analysis capabilities in GIS. He uses the term *spatial analysis method* in a broad sense to define a large class of complex methods to interpret spatial data and obtain decision-relevant information. Back in 1991, his assessment was as follows: "The geographical information revolution demands a new style of spatial analysis that is GIS appropriate and GIS proof. The existing spatial analytical toolbox is largely inadequate, consequently there is an urgent need to create more relevant methods and also to educate users not to expect the impossible when analysing geographical data. The real challenge is the need to develop new, largely automated, spatial data exploratory techniques that can cope with the nature of both the geographical data created by GIS and the skill base of typical GIS users."

Since then, the situation has improved considerably, and most major GIS vendors have added advanced analysis tools to their products. They have done so in two different ways: either by integrating more complex functionalities directly into the GIS, or by offering interfaces to special-purpose analytical software from third-party vendors. Given the current trend towards "open" GIS (cf. Sect. 3.1.3), the second option is becoming increasingly relevant. Scripting languages, such as ARC/INFO's AML (ARC Macro Language) or ArcView's[1] Avenue, serve to access and combine the various functionalities. In some cases, it is also possible to combine GIS and analysis functionalities by calling the required modules from some standard programming environment (such as a C++ program) via application programming interfaces (APIs). ESRI's Open Development Environment (ODE), for example, provides such an API to selected ARC/INFO functionalities.

In the remainder of this section, we briefly describe some of those analysis features that seem most relevant for environmental applications.

### 4.3.1 Spatial Operators

An obvious way to improve GIS analysis capabilities is based on more complex spatial operators. In particular, the combination of the distance function with spatial searches (cf. Sect. 3.2.1) yields a powerful set of analysis tools. Typical related queries are:

---

[1] ArcView is another GIS product of ESRI Inc. and a close relative of ARC/INFO.

- Find all yellow houses that are less than 2 km away from Lake Powell.
- Find all hospitals that are more than 3 km away from a bus route.
- Find the firestation closest to my house.

The first two of these queries involve the construction of a *buffer zone*, i.e., a polygon with a possibly complex shape that contains all points whose distance from the given polyline of polygon is less (or more) than a certain distance. The buffer zone then serves as input to a region query (cf. Sect. 3.3). The last query is a nearest-neighbor query that relates a single spatial object (here: the house) to a set of spatial objects (all firestations) to find the one with minimum distance. All queries involve computational geometry algorithms of considerable complexity. If they are applied in a database context, spatial join techniques may be used to optimize their performance.

A related class of analysis tools is based on the spatial modeling of three- or higher-dimensional objects. Three-dimensional modeling has traditionally been an important prerequisite for many applications in the geosciences. Set and similarity operators (cf. again Sect. 3.2.1) need to be computed efficiently in this context, as do more specific operators, such as the generation and manipulation of spline-based surfaces. Raper and Kelk [RK91] give an overview of related techniques. Higher-dimensional modeling is also increasingly used to simulate environmental phenomena, e.g., by three-dimensional transport models, or by spatio-temporal models, where one of the dimensions is used to model time [Lan92].

### 4.3.2 Optimization

Other analysis features establish a connection between GIS and optimization and decision support software. Such connections are becoming increasingly important, as environmental management is taking greater advantage of modern decision support technologies (see, for example, [BT97]).

Depending on historical reasons and marketing considerations, the decision support software is sometimes packaged as a separate system component, sometimes integrated into the GIS. ESRI, for example, offers a special network analysis package for its ArcView GIS. Their ArcView Network Analyst solves shortest-path and similar graph problems. Siemens Nixdorf, on the other hand, has long specialized in utilities applications, and many of its generic network analysis functionalities are consequently an integral part of its SICAD GIS. In addition, SICAD offers application-specific customizations for electricity networks (SICAD-UT-E), gas networks (SICAD-UT-G), and water networks (SICAD-UT-W).

### 4.3.3 Statistics and Visualization

While visualization is generally considered a core functionality of GIS, it should nevertheless be noted that most GIS visualization components concentrate on the classical map paradigm. What one sees on the screen is a map enhanced with some theme-specific labeling. The labels may be rather complex depending on the given application. The integration of business graphics, such as pie charts or histograms, for example, can be handled by most major commercial systems. However, if one compares these facilities to the visualization tools known to the statistics community, their deficiencies become clear.

Statistical packages provide advanced features to project and visualize multivariate data in a variety of ways that allow users to quickly recognize dependencies between variables. Certain dynamic techniques give users a sense of the overall shape of the data. Clusters and nonlinearities can easily be identified during such an *exploratory data analysis (EDA)*. GIS users could take advantage of such advanced analytical features in a large variety of applications that involve the interpretation of large spatial data sets. Nevertheless, classical GIS provide no such features.

Haining et al. [HMW96] describe the state of affairs as of 1996: "In order to carry out a program of SSA [Spatial Statistical Analysis], the user needs to access software that can store and manipulate spatially referenced information, can execute appropriate statistical analysis and finally allow good interactive visualization of the raw data or analysis outcomes. In each case there already exist packages which provide these facilities – in the order listed above GIS, statistical packages [...], and visualization packages. However, no current package provides all three."

In the meantime, more researchers and vendors have recognized the need for a consistent integration of GIS, statistics, and visualization. Most of the resulting system architectures leave the statistical computing and visualization to special-purpose software and construct interfaces between such packages and GIS to produce the desired results. This approach, which represents the current trend towards "open systems," should be seen in contrast to the assumption of some authors that GIS, statistical computing, and visualization are soon going to be integrated into comprehensive scientific computing environments. Rhyne, for example, predicts that the integration of GIS and visualization software will progress quickly along the following four phases: rudimentary (minimal data sharing), operational (consistency of data), functional (transparent communication), and merged (comprehensive toolkit) [Rhy97].

Symanzik et al. [SMCM97, CSMC97] used the Remote Procedure Call (RPC) protocol to construct a bidirectional interface between ESRI's ArcView GIS and XGobi [SCB91], an interactive dynamic statistical graphics program. In the terminology of Rhyne, their solution would be somewhere between operational and functional. Both ArcView and XGobi are set up to act as both RPC clients and RPC servers. This allows ArcView to pass data to an XGobi process to perform specific operations, and vice versa. In extension of this work, Symanzik et al. [SKS+97] later constructed a link between their ArcView/XGobi environment and the statistical computing environment XploRe [HKT95]. Symanzik et al. give an overview of the implementation (which is again RPC-based) and discuss an interesting application in satellite imagery interpretation.

Scott [Sco94] describes a link between ArcView and STATA, and Anselin and Bao [AB96] report on an interface between ArcView and the spatial data analysis software SpaceStat. ARC/INFO has been linked to MathSoft's S-PLUS [Mat96] and other statistical packages [HMW96].

### 4.3.4 Modeling

A last class of advanced analysis features is based on simulation models. Like statistical software, simulation software is rarely integrated into a GIS directly. One rather relies on interfaces that allow the simulation software to read and write geodata from the GIS. ARC/INFO, for example, maintains a variety of such interfaces, including one to FEFLOW, a simulation system for the modeling of transport phenomena. FEFLOW, which is marketed by a vendor called WASY [WAS97], has been developed independently of ARC/INFO over several years and provides advanced functionalities for the two- and three-dimensional modeling of groundwater flow. Like WASY and ESRI, many vendors of simulation software have close working relationships with GIS vendors to ensure the continued compatibility and efficiency of their interfaces.

Two collections of articles edited by Goodchild et al. [GPS93, GSP96] give a broad overview of related system solutions and developments.

## 4.4 Environmental Information Online

As noted in Chap. 2, environmental information is increasingly available in digital form. Classical online databases have been available since the early 1980s. Typical contents include fact sheets about chemical substances, information about recycling options, environmentally relevant standards and legislation, or surveys about ongoing research projects and publications. The databases are accessed primarily through special providers and dial-up connections. Users are usually charged a combination of a monthly or yearly base fee and usage-dependent charges. The traditional user community of such databases consists of researchers and practitioners working on environmental issues. Usage by non-experts or the general public has been rare.

As in many other domains, the rise of the World Wide Web has changed this situation fundamentally. Online database providers offer Web-based access to their systems, which has resulted in a considerable enlargement of their traditional user communities. Moreover, the ease with which data can be made available on the Web has increased both the volume and the quality of the data that is available. Many agencies and commercial vendors now offer their data directly on the Web without going through a third-party provider. Advanced functionalities that support the automatic conversion of legacy data sets into Web-compatible formats accelerate this trend. This applies not only to data in traditional office formats (such as MS Word) but also to non-standard data – such as maps.

We continue with a more detailed treatment of online databases, followed by a discussion of the Web and its impact on environmental information management.

### 4.4.1 Online Databases

Online databases have been a valuable source of up-to-date information on selected topics since the early 1980s. Environmental scientists have traditionally made heavy use of this medium, both for publishing and for retrieving information [Sto91, Gay91, GS92, Han92].

Access to online databases requires users to register with some service provider, in this context called *host*. The host bills its users periodically. Well-known hosts include Dialog (http://www.dialog.com), Data-Star (http://www.krinfo.com/dialog/publications/data-star-mini-catalogue.html), and STN International (http://www.cas.org/stn.html). Dialog and Data-Star belong to the American media company Knight-Ridder, and STN is run by a consortium consisting of the American Chemical Society (ACS), the German host FIZ, and the Japan Science and Technology Corporation (JICST).

Note that the host is usually different from the actual information provider. Typical information providers are scientific or economic institutions, or service organizations close to them. The Institute for Scientific Organization (ISI), for example, (http://www.isinet.com) collects data from many sources, reorganizes them, supplies keywords and thesaurus descriptors, and offers the resulting online databases to different hosts for distribution. Their *Current Contents* cover a wide range of topics and are offered by numerous hosts.

Online databases store their data in a record structure with both textual and numeric attributes. Attribute names mostly consist of two letters. Dialog, for example, uses the following acronyms for its most common attributes:

AB:   abstract
TI:   title
AU:   author
DT:   document type
DE:   descriptor
ID:   identifier (for classification purposes)
LA:   language
PY:   year of publication

There have been several attempts to define a taxonomy of online databases [Ort95, VB95], depending on their formats and contents. One can basically distinguish three types of databases:

1. *Bibliographic databases* contain catalog entries and possibly abstracts of publications related to a given topic. They closely resemble library catalogs except that their search capacities have traditionally been more powerful. Figure 4.6 gives an example from the database Enviroline. The first two columns contain the attribute names. Enviroline is offered by Dialog, Data-Star, and several other hosts (see, for example, http://www.kr-info.com/dialog/databases/html2.0/bl031.html). It corresponds to the print medium Environment Abstracts and covers a broad range of environmentally relevant information. For more information about Enviroline, see Figs. 4.6 and 4.7.

```
FN- DIALOG(R)File  40:Enviroline(R)|
CZ- (c) 1995 CIS, Inc. All rts. reserv.|
AN- 00247706
AA-(ENVIROLINE) 93-06159
TI- Toxicity Testing of Sediment Elutrates Based on Inhibition of
    Alpha-Glucosidase Biosynthesis in +i Bacillus licheniformis+r
AU- Campbell, Marjorie Univ of Florida, Gainesville; Bitton,
    Gabriel; Koopman, Ben
CS- Campbell, Marjorie Univ of Florida, Gainesville; Bitton,
    Gabriel; Koopman, Ben
JN- Arch Environ Contam Toxicol
PD- May 93
SO- v24, n4, P469(4)
LA- English
AV- Full text available from Congressional Information Service.
DT- research article
SF- 2 graph(s); 12 reference(s); 3 table(s)
AB- Elutriates were prepared from 66 sediments collected from nine
    hazardous-waste sites in Florida and screened by the Microtox
    and (gr)a-glucosidase biosynthesis (AGB) assays, using +i
    Photobacterium phosphoreum +r and +i Bacillus licheniformis+r,
    respectively. Total concentrations of lead, cadmium, zinc, and
    copper were determined for each elutrate. A linear relationship
    between AGTB and Microtox results was found. The percent
    agreement between the bwo bioassays was 85%. In terms of the
    heavy metals, AGB was more sensitive than Microtox results.
DE- (MAJOR) BIOASSAY; SEDIMENT; MEASUREMENTS & SENSING
DE- (MAJOR) METAL CONTAMINATION; LEAD; CADMIUM; ZINC; COPPER;
    TOXICOLOGY; BIOLOGICAL INDICATORS
DE- (MINOR) TOXIC SUBSTANCES
SH- 02
```

**Fig. 4.6.** Enviroline data record [Ort95]

2. *Literature databases*, also called *full-text databases*, go one step further by offering the complete text of selected publications. Reuters' Textline database, offered by Data-Star, is a typical example. Traditionally, full-

text databases were restricted to ASCII formats and could therefore not include graphics. Due to advanced communication and multimedia facilities, however, this is changing rapidly.

3. *Factual databases* do not refer to publications like the two previous types of online databases but rather contain original data sets. Typical contents include measurement data, physical and chemical properties of substances, project data, economic indices, or information about companies and institutions. We give three concrete examples.

   – The database PDLCOM offers manufacturer's and literature test data on the chemical compatibility and environmental stress crack resistance of plastics. PDLCOM is supplied by a company called Chemical Abstracts Service and offered by STN. PDLCOM can be reached on the Internet at http://info.cas.org/ONLINE/DBSS/pdlcomss.html.

   – The research database UFORDAT is maintained by the German Federal Environmental Agency and offered by several hosts including STN (http://info.cas.org/ONLINE/DBSS/ufordatss.html). UFORDAT stores information about environmentally relevant research projects in Germany and beyond.

   – CHEMTOX is a database containing information about substances, such as molecular formula and weight, chemical name, boiling and melting point, and toxicity. CHEMTOX is offered by a variety of hosts and can be reached, for example, at http://www.krinfo.com/dialog/databases/html2.0/bl0337.html.

Requests to online databases are formulated in proprietary query languages that allows users to specify the desired database and to formulate conditions on the data sets to be retrieved. Similar to SQL, this includes the formulation of basic boolean predicates on the attributes, using the two-letter acronyms discussed above. One can also specify the desired output format. Example query languages are Dialog, DSO, and Messenger. Note that both the query language and the user interface are host-specific.

Many databases offer special tools to support searches, such as indices or thesauri. The *basic index* of a database contains all terms occuring in selected text attributes. Depending on the kind of database, this may only concern special attributes (such as title or abstract) or the whole data set. Users can specify a number of terms to be matched against the basic index. The result of such a query consists of all records where any of the specified terms occurs in one of the text attributes. A *thesaurus* is a controlled domain-specific vocabulary that serves to capture the essential concepts of the chosen domain in a systematical manner. Each data set is indexed using one or more of the thesaurus terms (the *descriptors*). A thesaurus also maintains semantic associations between descriptors, such as synonym, generalization, and specialization relationships. These associations can often be used to improve search performance.

The contents of an online database are described in a structured description called a *bluesheet*. The bluesheet contains information about the database's thematic focus, the data sources, the period to which the data refers, the update cycle, and facts about the information provider. It also gives an example data set, an explanation of the attributes, an overview of search facilities (indices, thesaurus, etc.), and a listing of possible output formats. To find those databases that are relevant for a given search problem, most hosts offer a metadatabase with bluesheets of all of their databases.

Metadatabases from non-profit hosts often refer to a broader spectrum of online databases. The German Research Center for Environment and Health (GSF), for example, edits a *Metadatabase of Online Databases (DADB)* [VB93, BV95, VB95]. DADB concentrates on environmental chemicals and contains descriptions of more than 500 databases. It is complemented by metadatabases about Internet resources (DAIN), CD-ROMs (DACD), and printed documents (DALI).

Another important source of metainformation is the European Commission Host Organisation (ECHO). Their database *IM Guide* contains a large number of references to online databases and CD-ROMs. ECHO can be used free of charge and is available on the Internet at http://www.echo.lu. Figure 4.7 shows an example session where a user looks for data sources that relate to the search term *recycling*. The system finds 16 sources, including a CD-ROM called Environmental Solutions Database and the online database Enviroline.

Traditionally, queries were forwarded to online database via a dial-up connection to the host. Since the mid-1980s, Internet-based connections have become more common, starting with simple protocols such as FTP, Telnet, and Gopher. By now, many online databases offer comfortable Web interfaces based on the Hypertext Transfer Protocol (HTTP). Online databases were thus able to take advantage of the great popularization of the Internet. While there were almost exclusively used by specialists in the past, they now have the potential to become information providers for the mass market.

### 4.4.2 Environmental Information on the Web

The rise of the World Wide Web has greatly enhanced both the quality and the quantity of environmental data available. Large parts of these enhancements are offered to users practically without cost. With regard to quality, the World Wide Web offers excellent facilities for managing multimedia data, which is of obvious relevance for environmental applications. Regarding quantity, it is mainly the ease with which one can post information that causes the current exponential growth of available data.

Public agencies, for example, are increasingly recognizing the Web as a simple and cheap medium to distribute their data to other government agencies and the general public. Environmental agencies are often leading this movement; they usually have large data sets that are publically available in

```
?  f recycling
1.00  NUMBER OF HITS IS  16
?  s s=1;R=1 to 16
...
3.00/000005 ECHO: -IM GUIDE /COPYRIGHT ECHO
NA     : (ENVIRONMENTAL SOLUTIONS DATABASE) Environmental Solutions
         Database
AB     : Provides specific know-how and how-to solutions to
         industrial pollution problems. Covers water, wastewater,
         air, remediation, waste reduction, recycling and cost
         reduction.
CT     : ECOLOGY
PRTY   : FAC        ... Factual
ME     : CD         ... CD-ROM
UPD    : q          ... Every three month (Quarterly)
LA     : ENGL
...
3.00/000012 ECHO: -IM GUIDE /COPYRIGHT ECHO
NA     : (ENVIROLINE) Enviroline
AB     : ENVIROLINE is concerned with all aspects of environmental
         sciences. It corresponds to the monthly journal
         "Environment Abstracts". ENVIROLINE covers journal,
         magazine, newspaper and newsletter articles, conference
         proceedings and special reports. 70% of the documents are
         published within the US. It contains items on technical
         and policy oriented aspects of pollution, renewable and
         non-renewable resources, land use and misuse, use and abuse
         of marine environment, population planning and control,
         chemical, biological and radiation contamination,
         transport, waste treatment and disposal, weather
         modification and geophysical change, energy, food and drugs,
         wildlife, environmental education and environmental design.
CT     : ECOLOGY
DES    : specialized press; narcotic; demonstration; town planning;
         cleaning; grey literature; chemical pollution; pollution;
         united states of america; environment; demography; ecology;
         transport; energy; the study of the sea; politics; zoology;
         technology; sanitary engineering; meteorology; agro
         alimentary; geophysics; education
ME     : ON         ... Online base
UPD    : m          ... Monthly
LA     : ENGL
DBPR   : (CONGRESSIONAL INFORMATION SERVICE) Congressional
         Information Service
HOST   : ORBIT/QUESTEL
DB     : (ENVI) Enviroline
HOST   : KNIGHT-RIDDER INFORMATION - DIALOG, GATEWAY: MCRI
DB     : (ENVIROLINE) Enviroline
HOST   : FIZ TECHNIK
DB     : (ENVIROLINE) Enviroline
...
```

**Fig. 4.7.** ECHO query for online databases about *recycling* [Ort95]

principle, and there is an increasing demand on the part of the public to review them. Other major providers of environmental data include GIS vendors and companies who sell raw data, such as satellite imagery or measurement series. Besides their regular products, they usually offer some promotional material free of charge. Finally there are universities and other institutions providing educational material, again mostly free of charge. The great majority of courses offered relates to geographic information systems.

This section is not meant to provide a comprehensive list of Web sites devoted to environmental information. The goal is rather to describe various uses of the Internet to manage and distribute environmental data, and to illustrate these concepts by a few examples. For more material, the reader may consult one of the many search engines available on the Internet. We also provide a list of all relevant URLs appearing in this book in the appendix.

The most obvious way to find environmentally relevant information on the Web is to use a general search engine, such as Altavista (http://altavista.digital.com) or Lycos (http://www.lycos.com). One will likely experience the notorious disadvantages of this type of search tool: depending on the specificity of the query, users are often overwhelmed by a large number of answers. Only a small number of the cited documents may actually correspond to the objectives of the user. Nevertheless, with a query that is reasonably specific one will at least obtain several starting points for a more targeted search.

Another approach is to use a catalog or metainformation system that is geared specifically to environmentally relevant material. Like generic search engines, many of these systems allow keyword-oriented access and the formulation of simple queries that include boolean operators (AND, OR, etc.). Some also provide a thesaurus. Examples include (all URLs have to be prefixed by http://):

- AlfaWeb Hazardous Substance Information System (www.iai.fzk.de/~weidemann/lfu/lfu.htm)
- ASK – Global Change Directory of Information Services (ask.gcdis.usgcrp.gov:8080)
- Brown is Green Resource Conservation Program (www.brown.edu/Departments/Brown_Is_Green)
- Catalogue of Data Sources (www.mu.niedersachsen.de/cds/webcds)
- Central European Environmental Data Request Facility (www.cedar.univie.ac.at)
- Cygnus Group – Integration of Environmental and Business Concepts (www.cygnus-group.com)
- DAIN – Internet Resources on Environmental Protection (dino.wiz.uni-kassel.de/dain.html)
- Earth Observing System Data and Information System (spsosun.gsfc.nasa.gov/EOSDIS_main.html)
- Earth Pages (starsky.hitc.com/earth/earth.html)

- EcoWeb (ecosys.drdr.virginia.edu/EcoWeb.html)
- Environmental News Network (www.enn.com)
- Envirolink (www.envirolink.org)
- European Commission Host Organisation (www.echo.lu)
- Global Change Master Directory (gcmd.gsfc.nasa.gov)
- Global Network for Environmental Technology (gnet.together.org)
- Government Information Locator Service (www.epa.gov/gils)
- Landsat Pathfinder (amazon.sr.unh.edu/pathfinder1/index.html)
- National Environmental Data Index Catalog (www.nedi.gov/NEDI-Catalog)
- National Environmental Satellite Data and Information Service (ns.noaa.gov/NESDIS/NESDIS_Home.html)
- Register of Ecological Models (dino.wiz.uni-kassel.de/ecobas.html)
- UDK Austria (udk.bmu.gv.at)
- WWW Virtual Library Environment (earthsystems.org/Environment.shtml).

While many of these systems have been developed specifically for the Web, others are Web versions of older metainformation systems, such as library catalogs. Most of the materials indexed by these older systems are either paper documents (books, articles, etc.) or material that is digital but not available on the Internet (such as CD-ROMs). As more and more data sources are available online, however, these systems gradually assume search engine functionalities. As a result, the differences between search engines, online catalogs, and metainformation systems are gradually disappearing. Chapter 5 discusses this trend in more detail and gives several examples of metainformation systems in practice.

More advanced search and presentation functionalities are available for specific types of data. As mentioned in Sect. 2.5, for example, a company called Earth Observation Sciences (EOS) offers a Catalogue and Browse System (CBS) to provide content-oriented access to selected aerial and satellite imagery (http://www.eos.co.uk). NASA will soon offer content-oriented Web access to its well-known Earth Observing System Data and Information System (EOSDIS), which contains large amounts of imagery from a variety of satellite systems [Uni98b, DPSB97].

Many GIS vendors have started to sell special products to support their users in publishing maps on the Web. ESRI's MapObjects Internet Map Server, for example, helps Web developers to integrate spatial data and GIS functionalities into Web sites [ESR97b]. The ArcView Internet Map Server focuses on a wider audience by supporting users to publish any kind of ArcView GIS information on the Web. Similar products from other vendors include MapInfo's ProServer, or Genasys's Web Broker.

Note that map servers are usually much more powerful than simple image servers. On the one hand, map servers deliver not only static maps that have been produced ahead of time and stored at the server site. They rather allow users to specify the required information and possibly some display parameters, then generate the desired map at request time. The map thus serves as a customized visualization of some underlying (spatial or non-spatial) data set. On the other hand, the maps delivered are usually not just images but interactive entities that can be used to specify further queries. For example, users may be able to mark a point or rectangle on these maps, thereby requesting more detailed information about the selected location or region (*logical zoom*).

All these efforts clearly indicate the high potential GIS vendors see in Web-related business. ESRI estimates that by mid-1998 up to 20 percent of GIS computing may be provided through Internet and Intranet services [ESR97b].

The formats delivered by tools from different vendors are not always compatible but users have started to work on conversion tools to improve interoperability. The Environment Protection Authority of the Australian state of New South Wales, for example, has developed Web server mapping software to overlay map layers from different Web sites into interactive maps for viewing on standard Web browsers (http://www.epa.nsw.gov.au/soe/maps). Both source map layers and output maps are held in the standard GIF format used throughout the Web. The approach could therefore be used to combine output from different proprietary Web map servers. With some further development of tools and standards, an organization would thus only need to publish a GIF map file and accompanying metadata file on their Web site and notify a map server of its existence, in order to link their map layer to the pool of existing layers for their geographic area. This would in particular require a standard format for the metadata files to accompany map images. Chapter 5 will discuss related problems in greater detail.

As for Web-based *courseware*, most of the available material is on GIS-related topics. There is a search engine available that allows specifically to search for online GIS courses (http://www.frw.ruu.nl/eurogi). Among the university offerings we find three particularly noteworthy:

- The U.S. National Center for Geographic Information and Analysis (NC-GIA) offers a well-known GIS Core Curriculum [Nat97] that is now entirely Web-based (http://www.ncgia.ucsb.edu/education/ed.html) and provides a comprehensive coverage of the field.
- UNIGIS is an international consortium of universities led by Manchester Metropolitan University in the UK. It provides GIS courses and various degrees for distant learners. More information is available at http://www.unigis.org.
- A consortium led by U.C. Berkeley offers access to its public-domain GIS GRASSLinks [Uni98a]. GRASSLinks is a raster-based GIS that offers most standard GIS functionalities, including visualization, map overlay, buffering, and metadata management.

Most GIS vendors also offer access to tutorial materials on their Web sites. In addition to online introductions to their systems, this may include access to proceedings of user conferences or links to problem-specific discussion groups. ESRI's Web site (http://www.esri.com) is a typical example. There are also several commercial Web sites that offer GIS-related videos (see, for example, http://www.amproductions.com/contentg.html)

## 4.5 Environmental Management Information Systems

Most applications presented in this book are taken from *public* environmental information systems, i.e., from environmental information systems designed for and managed by public administrations. There is a simple reason for that: most environmental information systems in practice are run by public administrations. Gradually, however, the private sector is recognizing the need for collecting and managing environmentally relevant information as well. Books by Denton [Den94] and Steger [Ste93] (in German) discuss the possible benefits of environmental management at great length. Denton's book covers in particular the well-known Pollution Prevention Pays (3P) program of the American company 3M. According to 3M, the program saved over $500 million since its inception in 1975.

This trend towards environmental (data) management is reinforced by legislation efforts in many industrial countries that oblige companies to provide detailed reports on those activities that may have a significant environmental impact. The British Standard Institute, for example, published their specification for environmental management systems (BS 7750) in 1992 [Bri92]. The European Union followed suit one year later with their regulation on eco-management and audit schemes (EMAS regulation) [Cou93]. Both standards require that companies collect and compile data about inputs and outputs and their possible impact on the environment. On the input side, the EMAS regulation lists in particular raw materials, energy, and water. On the output side, it mentions solid and liquid waste, as well as air emissions and noise.

But those legal regulations are not the only reason why companies contemplate stronger information system support in the environmental sector. Wicke et al. [WHSS92] distinguish between internal and external uses of environmental information.

*Internally*, environmental information helps to make the environmental impact of a company's production facilities and its products more transparent. This is likely to have a direct impact on the departments concerned with production and materials. They can use this information to look for specific ways to minimize resource utilization, waste, and emissions, and to improve recyclability of their products.. Indirectly, other departments are concerned as well. Accounting and auditing may use the new insights to improve their bookkeeping procedures in order to provide a better documentation of material flow, marketing may be inspired to propose other ways of packaging a product, and so on.

*Externally,* the environmental data is primarily used to fulfill legal requirements. In addition, companies are likely to volunteer some of that information to their business partners and customers for a variety of reasons. Insurers, for example, have shown an increasing interest in environmental matters since courts tend to weigh related aspects more heavily in liability lawsuits. The liability problem also concerns investors who may often prefer companies that take possible environmental hazards explicitly into account. For example, the question whether a piece of land contains hazardous waste may well decide over the profitability of a related investment. Some investors also have strong ideological reasons to prefer companies that have shown concern for the environment. Investment companies are therefore offering "green funds" that include only shares of companies that fulfill certain environmental standards. Environmental information may also be useful to outside suppliers and corporate customers to optimize their own environmental strategies. Finally, environmental information about production processes and products – if favorable – is increasingly used for marketing purposes. In some countries and regions, such as Germany or Northern California, the "environmental correctness" of a product has become an important selling point.

### 4.5.1 System Architectures

Many companies are reacting to these insights and developments by building or purchasing specialized software. These *environmental management information systems (EMIS)* (also called *computer-aided environmental information and management (CAEM) systems* [Hil95]) may be stand-alone programs, or they may be integrated with existing information systems in the enterprise. Unlike most public environmental information systems, the majority of such systems concentrate on the technosphere rather than the biosphere. Most of the data stored in EMIS is about human-made systems (such as plants, production processes, or waste) and on their impact on the environment.

Stand-alone systems are by their very nature limited in their functionalities. Most of them are simple report generators that collect the necessary input via form-based interfaces and produce printed reports to fulfill legal requirements. They are sometimes coupled with a dictionary component that gives access to legal texts, or to data about hazardous substances. More sophisticated systems also provide some database functionalities to manage data about the company's technical equipment, important deadlines, past measurements, and other kinds of information that may be environmentally relevant. While relatively cheap in terms of licensing fees, the operating expenses of such stand-alone systems may greatly exceed their purchase price. In particular, data input can become a time-consuming chore and therefore a major cost factor. In addition, employees may quickly become tired of typing in data that they know is already stored elsewhere, and therefore – consciously or subconsciously – boycott the system.

For those reasons, more sophisticated systems provide interfaces to the existing infrastructure within the enterprise. Most important are connections to accounting and auditing, and to production. Based on those two possible cornerstones of an environmental management information system, one can distinguish two lines of related research and development: *accounting-oriented* and *production-oriented* systems. This taxonomy is similar to proposals by Hilty and Rautenstrauch [HR95]. We discuss those two directions in turn.

### 4.5.2 Accounting-Oriented Systems

Accounting-based systems revolve around the idea of an ecological balance sheet or *ecobalance*. An ecobalance is a systematic listing of the inputs and outputs associated with a production process, a product, or some other complex system. There exist several textbooks on ecobalances, including a well-known introduction by Braunschweig and Müller-Wenk [BMW93] (in German). Denton [Den94] discusses the subject from a North American point of view in the broader context of environmental management.

Ecobalances exist at different granularities, starting with simple process balances to balances that model products, plants, or the whole enterprise. In theory, these higher-level balances should constitute exact consolidations of the lower-level process balances. Due to missing and inaccurate data, however, this is rarely the case in practice. Ecobalances of highly complex systems are usually based on data sets that are themselves aggregations or estimates of lower-level figures. They are typically used to give management a general impression of the state of affairs, to fulfill legal requirements, and for public relations purposes. By themselves they are rarely suitable to detect problem areas or to search for concrete optimization strategies. For those purposes, one will first have to obtain more detailed balances of the products or processes in question.

Ecobalances for products that consider not only production but also the subsequent consumption and recycling phases are known as *life cycle assessments (LCAs)*. LCAs are arguably the most important type of ecobalance, which is reflected in the fact that the terms *ecobalance* and *LCA* are sometimes used synonymously. There have been considerable research and development efforts in this area. In 1997 the European Union issued a European Standard for LCAs. This standard, which is based on an earlier proposal of the German Federal Environmental Agency [Umw92], forms an essential component of the ISO 14000 series of standards on environmental management.

The standard consists of four parts, labeled ISO 14040 through ISO 14043. ISO 14040 [Eur97a] describes the principals and framework of the standard and proposes a decomposition of LCAs into the following four phases. ISO 14041 through 14043 describe those steps in more detail.

1. *Definition of goals and scope*, which is described in ISO 14041 [Eur97b], means defining the scope and granularity of the ecobalance(s) to be constructed. One has to decide which processes or entities to consider, which data to use, and what accuracy to attain. This is typically a group decision making process. Apart from generic software tools to support cooperative work ("groupware"), there is no software available that targets this phase specifically.

2. During *inventory analysis*, also discussed in ISO 14041, one seeks to obtain a complete listing of the mass and energy flows relating to the system in question. Depending on the intended applications, other kinds of resources (such as land use or human resources) may be included as well. A problem to be addressed during this phase concerns the *co-product allocation*, i.e., the allocation of inputs and outputs in those cases where a process is associated with more than one product [Ped93]. There are several software tools available to support inventory analysis, including many by non-profit organizations.

3. *Impact assessment*, which is the subject of ISO 14042 [Eur97c], serves to describe and possibly quantify the enviromental effects that are associated with the inputs and outputs identified previously. This requires in particular detailed information about the medical and ecological effects of chemical substances. Typical software support provides comfortable access to relevant databases; see [VB95] and Sect. 4.4 for an overview of available system solutions.

4. *Interpretation of results* is discussed in ISO 14043 [Eur97d]. It involves the assignment of weights to different kinds of impact, in relation to the objectives of the study to be conducted. The goal is to obtain a relatively simple representation of the possible alternatives, such that decision makers can make sensible comparisons across domains and classes of environmental effects. Of course, weights are subjective; they may vary greatly dependent on the political and ideological views of the participants. Like phase 1, the interpretation of results is typically a group decision making process. Software support concentrates on groupware and visualization tools. A notable exception is the EXCEPT system, which uses knowledge-based techniques for environmental impact assessment and interpretation [Hüb92, CDHH94].

For more comprehensive listings of related software support, see the article on LCAs by Miettinen [Mie93], the survey articles by Hilty and Rautenstrauch [Hil95, HR95], and the book edited by Guariso and Page [GP94].

As noted in ISO 14040 [Eur97a], "LCA is one of several environmental management techniques (e.g., risk assessment, environmental performance evaluation, environmental auditing, and environmental impact assessment) and may not be the most appropriate technique to use in all situations. LCA typically does not address the economic and social aspects of a product."

Ecobalances in the broader sense are an appropriate instrument to perform ex-post analyses of environmental impact. They are in particular suitable to fulfill legal reporting requirements that do not ask for causes as long as the environmental effects of a process are below the given threshold values.

However, if one is searching for explanations and for ways to improve the current situation, even detailed ecobalances are rarely sufficient. Researchers have therefore started to work on a more comprehensive concept that models the flow of matter and energy throughout the enterprise. Such *flow management systems* capture not only the available data on inputs and outputs but also knowledge about the underlying physical and chemical processes. As pointed out by Hilty and Rautenstrauch [HR95], this allows interpolations to make up for missing data. In addition, one can go beyond ex-post analyses and perform model-based simulations. As discussed in Sect. 4.2, this leads to prognoses and scenarios about environmental effects of ongoing processes or intended actions, which is of obvious interest to decision makers. Flow management systems often contain sophisticated animation components to visualize selected flows and their development over time.

The EcoNet project, which is based on Petri nets, represents an early step in this direction [SGH94, Hil95]. Other related developments have been reported in various workshop proceedings [GP94, SS95, SHH+96, AGHR97].

### 4.5.3 Production-Oriented Systems

Industrial manufacturing processes are amenable to a wide range of strategies to minimize negative environmental effects. The core question concerns the production process itself and its support by computer-based *production planning and control (PPC) systems.*

Traditional PPC systems serve to optimize the efficiency of some given production process. Important objectives in this context include maximization of throughput and machine utilization, minimization of production times, and optimization of stockpile management. An *environmental PPC system* [HR94] pursues several additional objectives: it tries to minimize emissions and waste, and it may also consider possible recycling options for the output(s) of the production process. This may or may not be fully compatible with the traditional objectives of a PPC system. In any case, one has to modify the objective function of the original system such that the environmental effects of the production process in question are taken into account.

Other ways to integrate environmental aspects into the production process concern the design of products, in particular with respect to recycling. The decomposability of products and the recyclability of the substances contained in them are factors that need to be taken into account early on during product design. Computer-aided design (CAD) systems can be programmed to support such an environmental orientation, e.g., by interfaces to substance databases or by special visualization components that emphasize decomposition and recycling options for the product. One can also go one step further and use PPC systems to plan not only the production process but also the subsequent recycling process. Such approaches gain considerably in efficiency if they can share their data with the PPC systems used in production [Rau94, KR95, KSE95].

Clearly these ideas go beyond the reporting objectives that have been the main focus of ecobalances. Similarly to flow management systems, PPC-based approaches represent a *constructive* approach in the sense that they help not only to identify weak points in the enterprise but also to formulate alternative strategies to improve the current situation.

### 4.5.4 Commercial System Solutions

The market for environmental management information systems is growing quickly. Most major software vendors offer at least one system solution. In addition, there are more than 100 software packages offered by smaller companies, most of which cover only a small part of the functionalities discussed above. Software is highly country-specific and there are frequent updates to stay in accordance with current legislation. The technical focus of the available systems varies widely, depending on their different histories.

Several system products come from a health and safety background. IBM's Chemical Health and Safety Environment System (CHEMS), for example, has its traditional focus on chemical information management, chemical administration, and health and safety issues [Kra93]. In addition there are modules to perform internal environmental impact assessments, check legal compliance of certain measures, and perform environmental auditing tasks.

Other systems were developed from databases containing substance-related information. IGS care of Siemens Nixdorf, for example, is based on IGS, an information system for hazardous substances and related legal documents [LPS93]. BMW's ZEUS is a central information system about environmentally relevant substances [Ora94].

Yet other systems take a more generic approach by providing basic database functionalities (including a data model) and application-specific user interfaces for data input, data manipulation, and report generation. Both debis-UIS of CAP debis [CAP94] and Digital's DEC-BUIS [MWWW93] are such framework systems.

It is difficult to predict which way the market will go. There can be no doubt that the new ISO 14040 standards will have a major impact. Customers will demand compatibility, and vendors will have to follow suit. This should result in a more transparent market, where software packages will be easier to evaluate and compare. This would greatly benefit many consumers, who currently depend on expensive consulting services to evaluate software and to adapt it to their particular requirements. Another likely result ist that there will be some kind of selection process, which will drive the majority of current vendors out of the market. There will probably be a small number of vendors selling open framework systems, and a larger number of companies offering special-purpose components that fit into those frameworks. The framework systems may well come from companies that already have a major stake in the market for production or accounting software. Recent activities by SAP [HG96], Debis [CAP94], and other major software vendors document their continuous interest in the environmental market.

At this point, however, many potential customers still have major doubts about the cost/benefit ratio of environmental management information systems. Many of the advantages listed above are still somewhat uncertain and hard to quantify, although progress is being made in that respect [Den94, Vor97]. Possible savings in resource utilization and waste disposal are difficult to prove. The dangers that result from possible liability lawsuits are not considered a significant problem. In the United States, the probability of being found guilty is small. In Europe, on the other hand, recent legislation (such as the German Environmental Liability Law [AG94]) have made it easier for individuals to get a company convicted of environmental crimes. In contrast to United States levels, however, fines and damage awards are still relatively low.

As a result, most commercial system solutions concentrate on the fulfillment of concrete legal reporting requirements. There is little demand for more comprehensive solutions that not only focus on emissions but take material flow as a whole into account.

### 4.5.5 Outlook

The situation described above can be seen as analogous to the "end-of-pipe" solutions in other areas of environmental protection and management. Legislative support for more comprehensive solutions cannot be reasonably expected as long as many Western industrial economies are still in a fragile state. Until there are visible signs of a long-term recovery, legislators will be hesitant to impose stricter guidelines on companies. It is thus up to software vendors and researchers to show more convincingly that a comprehensive environmental (information) management has more to offer than marketing advantages and a slight drop in insurance premiums – that it can in fact save resources and improve the overall efficiency of production.

From a research point of view, the whole area of environmental management information systems is of very recent origin and still very much in flux. In addition, many related issues are of an institutional character and therefore depend on the given legislative and organizational framework. As a result, publications in this area are often specific to a particular country or branch of industry.

Readers interested in learning more about EMIS should consult some of the available survey articles and conference proceedings for more information. Hilty and Rautenstrauch [Hil95, HR95] give an overview of the field as of 1995. A later literature survey by the same authors [HR97] covers more recent publications in the field and structures them depending on their level of abstraction. Arndt and Günther [AG96] briefly describe the current legal and organizational situation in Europe and present possible system architectures that fit into this framework. The conference proceedings edited by Haasis et al. [HHKR95, HHH+95] give a good impression of ongoing projects. Except for [Hil95], all of these publications are in German. At least [HR97], however, should also be useful to readers who do not speak German, simply because it includes a large number of references to articles written in English. Some of the relevant work in North America is described in Denton's book on environmental management [Den94].

## 4.6 UIS Baden-Württemberg: An Integrated Public EIS

Baden-Württemberg is a highly industrialized state in the southwest of Germany with a population of about 10 million. The State Ministry of the Environment was founded in 1987, which was also the starting point for a large-scale project called *UIS Baden-Württemberg*. UIS is an acronym for the German translation of *environmental information system*. UIS was conceived as a showcase project for the state's environmental activities in the wake of environmental disasters like Chernobyl and the Sandoz Rhine pollution case. The state hired a major management consulting company to conceive a first system design in cooperation with various government agencies [UM87]. This design has since been implemented and developed further under the leadership of a working group in the Ministry of the Environment (since 1996 the Ministry of Environment and Traffic). The annual budget of the UIS project group is between 30 and 40 million DM per year.

A good overview of the complete system as of 1993 is given in a special chapter of [JKPR93]. Since then, the main focus has been to adapt UIS to open client/server architectures, WWW technology, and middleware standards such as CORBA [MSS96, May97]. UIS can be accessed via the Internet at http://www.uis-extern.um.bwl.de.

### 4.6.1 Objectives and System Architecture

The main objectives of UIS are [May93, MSS96, May97]:

- to support the administration in their environmental management and planning tasks;
- to implement an efficient environmental monitoring, including data capture, analysis, and forecasting;
- to support the management of environmental emergencies;
- to make environmental information available to the executive branch as well as to the general public;
- to protect past investments by coordinating existing system solutions and integrating them into a common system architecture.

From the start, UIS was not conceived as a monolithic system but as a network of subsystems and services. It was particularly important to be able to integrate existing systems, in order to avoid losing past investments and to ensure acceptance among other government agencies that are involved in environmentally relevant tasks. TCP/IP, relational databases, and client/server technology had just become popular at the time UIS was conceived, and have since become cornerstones of the overall system architecture. The World Wide Web was integrated into the design soon after it had become widely available. More recent system components follow middleware architectures with *object request brokers (ORBs)* negotiating between UIS service providers and consumers [KKT+96, RMW97].

While the UIS project has been a major force for the harmonization of terminologies and naming conventions throughout the state, constructing an all-encompassing data model (e.g., by means of an entity-relationship approach) was not attempted. This decision results from the lessons learnt in the 1970s and 1980s when many companies attempted – and failed – to build an enterprise-wide data model. UIS adopts a bottom-up approach instead, building on existing application-specific data models and database schemata. Thesauri and metainformation systems (such as the UDK, cf. Chap. 5) are used to establish cross-references and allow users to access subsystems with their particular terminologies.

The original UIS design [UM87] has gone through a natural evolution over the years. Some of the original objectives and modules have become more relevant than expected, and vice versa. Today one can distinguish five kinds of system components (Fig. 4.8):

1. *Generic base components* are systems that are not exclusively used for environmental tasks but for administrative tasks in general. These systems are part of the government's general computing infrastructure. Typical examples are official digital map collections, statistical databases, the cadastre, or the government intranet.
2. *UIS-specific base components* extend this basic infrastructure. Typical examples are data dictionaries, regulations for the formatting and exchange of spatial data sets, or operative rules for measuring networks.
3. *Task-specific components* are systems that have been designed for particular environmental applications. They are the most important source of environmental data and provide an essential foundation for UIS. Many task-specific components are concerned with capture and aggregation of environmental raw data, such as measuring series or aerial imagery. Others are devoted to more analytical tasks like water treatment monitoring.
4. *Integrative components* are intended for inter-agency and inter-domain aggregation of environmental data. The goal of these systems is to provide an integrated view of complex environmental phenomena that transcend the traditional boundaries defined by media and organizational hierarchies.

5. The *environment and traffic information service UVIS* is a strategic management information system for high-level decision support. It provides compact visualizations of complex, highly aggregated data from numerous connected subsystems.

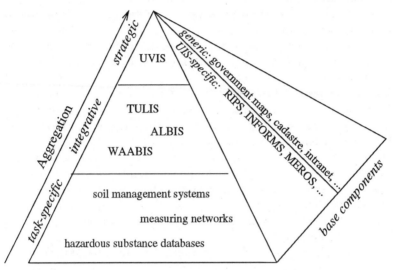

**Fig. 4.8.** Pyramid architecture of the UIS Baden-Württemberg

Generic base components are not specific to environmental applications and therefore outside the scope of this book. In the following, we discuss the remaining four classes of UIS system components in turn.

### 4.6.2 UIS-Specific Base Components

Besides a solid administrative and technical infrastructure, which is provided by the generic base components, a system like UIS requires a consistent infrastructure of data formats and processing guidelines. UIS-specific base components serve to provide such an infrastructure by standardizing the exchange and interpretation of environmental data sets. Among the most important UIS-specific base components are the systems INFORMS, MEROS, and RIPS.

*INFORMS* is a comprehensive UIS data dictionary [UD92]. It contains (meta-)data about available data sets, such as the quality of the data, the responsible agency, and the technical and organizational conditions for accessing the data.

*MEROS (Measurement Series Management System)* and *RIPS (Spatial Information and Planning System)* both have a more specific focus. MEROS collects data sets from a variety of measuring networks and stores them (redundantly) in a central location. RIPS is the central spatial data management component of UIS [Mül93]. In their respective domains, each of those two systems provides access to a large part of the relevant data held by the state government.

Given that the data is both heterogenous and distributed (among numerous agencies), there are essentially three design options for a system like MEROS or RIPS.

The first possibility is to build a metainformation system that just contains reference information on what kind of information is available and how to obtain it. In other words, one would build a specialized data dictionary. The obvious drawback of this solution is that the user has to perform some effort to obtain the data, either via computer network or, still more common, via telephone and floppy disk. Formats may differ, depending on the preferences of the information provider. If the user (i.e., the information consumer) is trying to retrieve and synthesize data from different sources, this may cause major problems. The advantages of the metainformation architecture on the provider side mirror these problems on the consumer side. The effort needed to build and maintain such a system is comparatively small. Information providers can publish their data in formats they prefer. Heterogeneity problems are left to the information consumer to solve.

The second design alternative is to build a virtual database, based on the concept of a database view. Users can thus retrieve and possibly update data sets through a uniform and standardized access facility. The data sets, however, are not stored at some central location. They remain stored locally at the responsible government agency. Whenever a user requests a data set, it is automatically forwarded and reformatted according to the user specifications. This obviously requires a much greater effort on the part of the information provider and at the central system administration level. Information providers need to keep their data correct and accessible. System administrators need to make sure that communication and reformatting do not lead to unacceptable retrieval delays.

The third option is to build a central database. This is most comfortable for information consumers but requires again a major effort by information providers and at the system administration level. Most, if not all, of the information in such a database will be redundant because it will typically also be stored at the local or specialized agencies that capture the data. One consequently requires mechanisms to insure the currency and consistency of the different copies.

While integrating ideas from all three alternatives, MEROS and RIPS both tend to focus on the third option. They can be regarded as data warehouses in the sense of [Hal95] and [CD97]. To illustrate the relevant concepts, we now discuss RIPS in some more detail.

RIPS offers access to a central database of geographic data that consists of copies of local geographic data sets, including:

- vectorized administrative maps (state and county borders, etc.)
- digital elevation models
- rasterized topographic maps at scales between 1:25 000 and 1:500 000.

The data holdings are updated frequently. To provide comfortable access, RIPS offers a metadata component that serves as an entry point and that helps users to navigate in this spatial database. A RIPS method base contains algorithms for the most common spatial queries, such as point searches, range searches, and map overlay.

The system is implemented on the basis of two commercial products: the relational database system Oracle, and the object-oriented geographic information system Smallworld. All spatial data sets are stored in the object-oriented spatial data model Smallworld Datastore [The97, BN97], which is compatible with the storage format of the state cadastral administration. Non-spatial data, including the metadata, is stored in the Oracle relational database. Spatial and non-spatial data items are related to each other via unique object identifiers. In addition, the Oracle database maintains minimum bounding rectangles for all spatial objects to allow approximate spatial searches without consulting the Smallworld GIS for the actual object geometries (cf. the filter step described in Sect. 3.3). Interaction with the Oracle subsystem is mainly through SQL*FORMS, an Oracle graphical query tool.

RIPS data sets can easily be combined with each other and with state cadastral data. RIPS thus creates a common infrastructure to read and process spatial data throughout the state and beyond. This represents a major advance compared to previous system approaches that were highly proprietary and format-specific.

### 4.6.3 Task-Specific Components

Task-specific components support the capture and management of specific types of environmental raw data. Typical examples are:

- measuring networks and related software
- laboratory software for data capture and interpretation
- software systems for garbage management and soil protection
- food- and drug-related databases
- hazardous substance databases
- data collections on industrial sites.

Many of these systems had their origins long before UIS was introduced. They were typically developed and maintained under the supervision of some specialized task force or agency that was in charge of the data in question. Other task-specific components are of a more experimental nature. In Chap. 2 we already discussed two related prototype systems in more detail: WANDA uses techniques for the management of uncertainty to support the evaluation of water measurement data. RESEDA is a knowledge-based approach to the interpretation of satellite imagery.

Specialized computer systems for particular tasks and agencies have thus been common for a long time in environmental administration. What is new in UIS is a methodical approach to integrate those systems into a common framework.

### 4.6.4 Integrative Components

The function of the *integrative components* is to collect data from task-specific components throughout the state, perform domain- and application-specific aggregations on them, and present them in their spatial and temporal context. Simulation models and scenarios play an important role for this kind of aggregation. Typical examples of integrative components are the systems ALBIS, TULIS, and WAABIS.

*ALBIS (Species, Landscape and Biotope Information System)* is an information system about the biosphere, in particular about living species, natural resources, and biotopes. Its target group consists of mid-level administrators at the state and county governments. ALBIS receives data from local agencies responsible for natural resource management, in particular forestry and wild and game management. This data is then aggregated and prepared for visualization via graphics and tables. The presentation is supplemented by a multimedia component that allows one to see images and hear soundbites from selected regions or species.

*TULIS (Technosphere and Air Information System)* [Koh93] provides a direct counterpart to ALBIS. While ALBIS is concerned with data about the biosphere, TULIS manages data about the air and the *techno*sphere. TULIS has been designed to support mid-level administrators in tasks related to clean air monitoring and facility surveillance. The system is closely tied to a task-specific component that maintains current data on industrial sites that may have an impact on the environment. Based on the incoming data, TULIS computes middle- and long-term trends in emission and air pollution and identifies actual or potential conflicts with state laws and policies. The results are displayed as thematic maps: for example, one may produce a map showing the intersections between protected areas (national parks, groundwater collection areas, etc.) and incidences of high pollution.

*WAABIS* is the state's *Water, Waste, and Soil Information System*. WAABIS has been implemented jointly by the state government and the county administrations. It continues to rely on a tight integration of related subsystems at different levels of administration.

### 4.6.5 UVIS: The Management Information System

The *Environment and Traffic Information Service (UVIS)*[2] is an ambitious attempt to further aggregate and synthesize the available data in order to provide efficient decision support for top-level administrators. UVIS is tightly integrated with the integrative components described in the previous section as well as with selected task-specific components. Table 4.1 shows some of the data sets available under UVIS.

**Table 4.1.** Selected data sets available under UVIS [Hen93]

| Topic | Time Period |
|---|---|
| *Ground Measurements* | |
| lead | 1988 – present |
| cadmium | 1988 – present |
| *Demographic Data* | |
| population | 1970 – present |
| population density | 1970 – present |
| land utilization | 1981 – present |
| *Natural Resources* | |
| parks | 1990 – present |
| biotopes | 1990 – present |
| *Forestry* | |
| forest damage | 1985 – present |
| *Garbage* | |
| domestic garbage | 1989 – present |
| industrial garbage | 1977 – present |
| garbage collection and | 1977 – present |
| *Air* | |
| $SO_2$ emissions | 1978 – present |
| ozone | 1978 – present |
| precipitation | 1978 – present |
| *Groundwater* | |
| pesticides | 1989 – present |
| *Radioactivity* | |
| Geiger-Müller counters | 1989 – present |

In 1997 the UIS team presented the current version of UVIS, which is based on open client/server architecture with TCP/IP-based communication [HWS97, RMW97, RWGM97]. This system (originally called UFIS II) supports a further abstraction of UIS components as services that are activated by means of standard intranet technology, such as the Hypertext Markup Language (HTML), the Common Gateway Interface (CGI) [McC94], or Java applets [SUN98].

---

[2] From 1984 until 1997 this project was called UFIS, an acronym for the German translation of *Environmental Management Information System*. Note, however, that UFIS/UVIS is *not* an environmental management information system in the sense of Sect. 4.5. The terminology is inconsistent at that point.

Users interact with UVIS through a client consisting of a Web browser, a broker (called *PDN = Personal Data Node*), and various software tools for presentation purposes, including ArcView GIS and MS Office. Users specify their queries via a graphical user interface (Fig. 4.9). To improve the efficiency and selectivity of a query, they can consult a special metainformation system (the *locator*). The query is then forwarded to the relevant services for processing. Services may be located anywhere in the network. Typical UVIS output consists of a task-specific thematic map combined with tabular non-spatial data (Fig. 4.10).

To support session management, the UVIS designers developed a special query interpreter called *WebQuery*. WebQuery supports long, multi-step communication exchanges between a Web browser and a relational database and performs the required conversions.

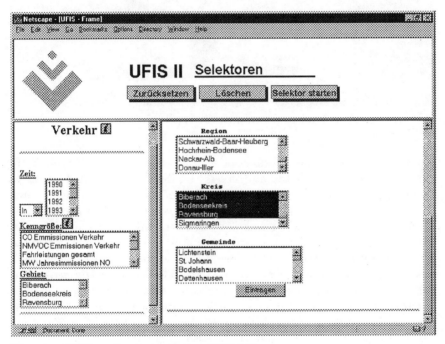

**Fig. 4.9.** UVIS application: selection [HWS97]

Additional query support in UVIS is provided by a locator, a thesaurus, and a gazetteer. The *locator* has already been mentioned: it is a metainformation system containing an index and a description of all information sources available under UVIS. The index can be searched by keywords, geographic references, temporal references, and information provider. A more advanced version of the locator is included in the *German Environmental Information Network (GEIN)*, which has been developed as part of the G-

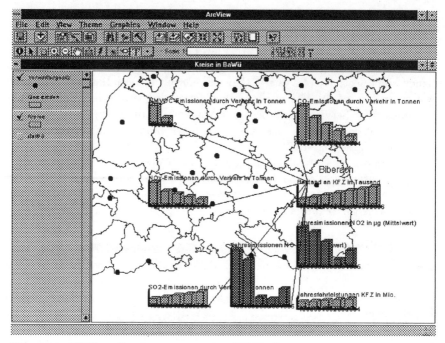

**Fig. 4.10.** UVIS application: presentation [HWS97]

7 initiative for a Global Information Infrastructure [SL96, RWGM97]. To improve the efficiency of keyword access, UVIS offers integrated access to the polyhierarchical environmental *thesaurus* of the German Federal Environmental Agency [Bat94]. The thesaurus offers a comprehensive collection of environmentally relevant terms and their semantic associations (in particular synonym, generalization, and specialization relationships). Users can then decide whether they want to include any of the related terms in addition to the keywords they actually specified. The *gazetteer* is a geographic index that allows users to specify either a rectangular region or a geographic or administrative name (such as *Rocky Mountains* or *Colorado*). Topological relationships between geographic entities (such as *includes* or *overlaps*) are represented in the gazetteer and can be used as query predicates.

### 4.6.6 Outlook

A major focus of the current UIS work is a reorientation towards object request broker architectures such as CORBA [Obj95a, Obj95b]. CORBA, the *Common Object Request Broker Architecture*, is recommended by a broad coalition of vendors and user groups as a platform for systems integration. It can be used in particular to encapsulate existing applications ("legacy wrapping") and to make them available as intranet (or Internet) services. CORBA and similar approaches thus help to bridge heterogeneities, to simplify the integration process between existing system components, and to improve the overall quality of service.

Several research projects have already shown the considerable potential of object request broker technologies for environmental and scientific data management [MMS+96]. Both Baden-Württemberg and Brandenburg, a state in the northeast of Germany, have reported encouraging results with CORBA-based system architectures for their environmental information systems [SSW96, KKT+96, RMW97]. The UIS group is currently using CORBA/IDL (Interface Definition Language) to implement a registry of available services [MSS96, May97]. CORBA is also under consideration by the designers of the European Catalogue of Data Sources (cf. Sect. 5.3).

## 4.7 Summary

This chapter was devoted to data analysis. In this context, the term data analysis refers to the preparation of information to fulfill specific requirements of decision makers.

We began with a discussion of environmental monitoring (Sect. 4.1). Environmental monitoring is a particularly important application of data analysis, as it requires sophisticated tools to identify events and trends that are potentially dangerous.

We continued with a description of the most important classes of analysis techniques. Section 4.2 gave an overview of simulation models for environmental applications. In Sect. 4.3 we listed some of the data analysis tools offered by commercial geographic information systems. Section 4.4 discussed online databases and the impact of the World Wide Web on environmental information management, and Sect. 4.5 gave an overview of environmental information systems in the enterprise. Section 4.6 concluded with a case study: UIS is a comprehensive approach to environmental information management that uses many of the analysis techniques presented in this chapter.

# 5. Metadata Management

As we saw in the previous chapters, there is a major need for convenient navigation aids that help users to take advantage of network-based, distributed information, regardless of their computer literacy. Starting from an environmental query or problem formulation, such navigation aids should help users to localize the relevant data sets and to retrieve them quickly and in a user-friendly manner.

An essential prerequisite for both navigation and data transfer is the availability of appropriate *metadata*, i.e., data about the format and the contents of the data. The key idea is to enhance data sets by concise self-descriptions in order to improve both the speed and the accuracy of related search operations. The metadata serves as a kind of online documentation that can be read and utilized by appropriate tools as well as by human users. Note that there is no intrinsic distinction between data and metadata; it is rather a question of context whether a given data item represents metadata or not.

In this chapter we discuss the question of metadata in environmental data management in greater detail. Section 5.1 gives a more elaborate definition of metadata and shows how metadata can be integrated into a traditional data management architecture. Sections 5.2–5.4 describe several concrete approaches to metadata management. Section 5.2 presents the U.S. initiative to create a National Information Infrastructure (NII) and, within the NII framework, a National Spatial Data Infrastructure (NSDI). This includes discussions of the Government Information Locator Service (GILS), the Spatial Data Transfer Standard (SDTS), and the FGDC Content Standards for Digital Geospatial Metadata. Sections 5.3 and 5.4 continue with descriptions of two European systems: the Catalogue of Data Sources (CDS) developed by the European Environment Agency (EEA), and an environmental data catalog called UDK, whose development was coordinated by Austria and several German state governments. Section 5.5 concludes with a summary and an outlook on future work.

## 5.1 Metadata and Data Modeling

Our further discussion is based on the three-way data model described in Chap. 1. We distinguish between environmental objects, environmental data objects, and environmental metadata (Fig. 1.1). As we already discussed at length, the data flow in many environmental applications is usually associated with a complex aggregation process to provide decision support at various levels of responsibility. It closely resembles the data flow known from classical business applications: data capture, data storage, and data analysis.

Metadata may be collected throughout this aggregation process and built into the corresponding data structures. As Strebel et al. [SMN94] have pointed out, it is important to collect metadata in a timely manner – if possible, simultaneously with the collection of the original data. Figure 5.1 pictures their empirical results. It shows on the one hand (dotted line) how the effort to collect metadata increases dramatically if the collection effort is delayed with respect to the original data capture. On the other hand (solid line), the amount of metadata that can typically be recovered drops with increasing delay.

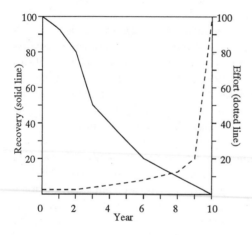

**Fig. 5.1.** Effect of delay between data capture and metadata collection [SMN94]

So far, metadata collection is a mostly manual process. The problem is that the most qualified people to perform this process are the data providers themselves. Those people, however, usually have little motivation to do so – they know how to obtain and interpret the data. What we see here is a variation of the classical software documentation problem. In analogy, there are two solutions: one can either establish an organizational framework that requires data providers to perform the necessary metadata capture, or one can search for ways to extract the metadata automatically.

The first option is typical for hierarchical organizations, such as companies and government agencies. To ensure the continuity and quality of the metadata, and in order to maintain motivation among employees, it is important to complement such a policy by software tools that facilitate the extraction process. Web browsers with their form-based graphical interfaces offer a solution that is both cost-efficient and simple to implement.

The automatic extraction of metadata is slowly becoming a serious alternative [DY97]. Riekert et al. [RMW97] present an approach where incoming documents are matched against a thesaurus to extract relevant terms. They intend to add a similar matching procedure for a gazetteer, i.e., a geographic index (cf. Sect. 4.6.5). In addition to matching geographic names, this requires a spatial component that determines the spatial extension the data set refers to.

Once the metadata has been collected, it can be used for a variety of purposes, especially during the data analysis phase:

- Computerized environmental information systems are able to collect and process much greater amounts of data than anybody would have imagined only a few years ago. Automatic data capture and measurement collects terabytes of new data per day [CC90]. Even in processed form, this kind of data is impossible to browse manually in order to find the information relevant for a given task. Modern information retrieval tools allow automatic or semi-automatic *filtering* of the available data in order to find quickly those data sets one is looking for. Metadata forms an important foundation for these tools by serving as a condensed representation of the underlying data. As such, it supports browsing, navigation, and content-oriented indexing.
- Environmental data management is extremely *heterogenous*, both in terms of hardware and software platforms. Data is organized according to a wide variety of data models, depending on the primary objectives of the particular agency in charge. Metadata can help to overcome these heterogeneities by specifying the platforms on which a given data item is located. This way, appropriate conversion routines can be introduced (semi-)automatically, wherever necessary.
- Environmental data is frequently *uncertain*. Metadata can be used to specify the accuracy of a data item, so users can judge from the metadata whether the corresponding environmental data objects are relevant for their current needs.
- Metadata can also help to *inventory* existing data holdings, to unify naming schemes, and to record relationships between different data items and data sets. This aspect of metadata has become very popular as one of the core functionalities of *data warehouses* [Hal95, CD97].

The concept of metadata is not new. Online documentation of programs and data sets has been in common use for many years. As we saw in Sect. 4.4.1, online database providers have long offered index databases that help users to find the data sources relevant for a given query. Machine-readable metadata has also been known for a long time, in particular in the context of relational databases, where the internal database structure (the *database schema*) is typically represented in a relational format itself. What is new, is the more systematic approach to providing machine-readable metadata, and the trend to standardize metadata in certain application areas.

For the subsequent discussion, it is useful to distinguish between two kinds of metadata [MDF95]. The term *denotative metadata* is used to refer to the kind of metadata that describes the *logical structure* of a data set; a relational schema would be a typical example. The term *annotative metadata*, on the other hand, is used to describe data that provides *content-oriented context information*, such as the documentation of the measuring series described above. Following Melton et al. [MDF95], further examples of annotative metadata include "information in scientific notebooks, instrument logs, manuals, and reports that document the platform and instrument conditions, the operational environment, interfering sources of noise, and that uniquely identify the software and computer platforms used for analysis, modeling and simulation." In the remainder of this chapter, we will concentrate on annotative metadata and use the term *metadata* in that sense, unless noted otherwise.

The relevance of metadata for the management and analysis of complex data sets was pointed out early on by McCarthy [McC82] and pursued further in the area of statistical and scientific databases. Siegel and Madnick [SM91] built on those ideas, concentrating on possible applications in financial data analysis. The IEEE Mass Storage Systems and Technology Committee has sponsored several metadata workshops whose results are available on the Web (URL http://www.llnl.gov/liv_comp/metadata/metadata.html).

The use of metadata in geographic and environmental information systems is of a more recent nature [Rad91]. Lately, however, there has been broad agreement that metadata are a crucial factor to improve both the quality and the availability of geographic and environmental data. Several conferences on spatial databases and GIS have devoted parts of their program to metadata [GS91, ESR91, ESR95b], and there has been a variety of workshops dedicated exclusively to metadata management in the geosciences and environmental sciences [MSNRW91, MDF95].

In terms of practical consequences, metadata technology is increasingly being integrated into commercial GIS. Most commercial systems have always maintained some basic metadata on the objects to be administered. ARC/INFO, for example, generates and maintains metadata on the spatial registration, projection, and tolerances of a coverage or grid [ESR95a]. Every time one creates a coverage, the system creates a set of metadata files, including the TIC file (containing data about the coverage's coordinate registration), the LOG file (tracking all ARC operations performed on the coverage), and the BND file (containing the coordinate values that denote the outer boundary or spatial extent of your coverage). There is also denotative metadata giving some schema information on the INFO tables that contain the non-spatial data components.

The practical use of metadata, however, extends far beyond this somewhat narrow scope. One trend is to collect more information about the detailed content of the data. Vendors typically choose some bibliography-style format to represent this information. In the case of spatial data, conformity with the FGDC Content Standards (see Sect. 5.2.3) is increasingly required. The ARC/INFO component DOCUMENT.AML [ESR95a] is a typical example of such a tool.

Another trend is to describe the history and quality (also called *lineage*) of data sets and their sources in more detail. Geolineus from Geographic Designs is a common tool for this purpose [Geo95]. Geolineus represents the data in a GIS by means of dataflow diagrams, where coverages and grids are shown as icons. Icons along the top of the diagram represent the *source data* on which the GIS is based. Icons further down represent data layers that were *derived* with spatial analysis operations like *buffer* or *intersect*. Finally, icons at the bottom of the diagram represent *products*, i.e., derived data items that represent the final steps in a GIS application. Geolineus shows the type of data in the corresponding layer for each icon and maintains command histories for each coverage. The system allows documentation about each layer to be stored in a frame-based format.

Drew and Ying describe a concrete approach to use metadata in order to provide uniform access to a heterogenous collection of GIS and spatial databases [DY97]. Based on metadata about those systems and their contents, their GeoChange system serves as a navigation and access tool. To a large extent, it is non-intrusive, i.e., it can be implemented on top of an existing collection of independent systems without major changes to the underlying architectures and implementations.

Other trends in metadata management include the inclusion of more spatial elements in the metadata itself [Sea95] and the use of metadata to describe and access not only other data sets, but also models and algorithms. We already described the Register of Ecological Models (REM) in Sect. 4.2.2. A related effort is described by Lenz et al. [LKH+94].

Parallel to these application developments, metadata management has become a focus in an increasing number of government R&D projects. Besides the efforts described in the following sections, there has been a project by the European Space Agency (ESA) to develop an online geosciences metadata system, called the ESA Prototype International Directory [Wal91]. At about the same time, the United Nations Environmental Program (UNEP) started its project on Harmonization of Environmental Measurements (HEM, cf. Sect. 4.2.3) [KMB91]. This was later followed by UNEP's Global Resource Information Database (GRID), which includes a Metadata Directory (GRID MdD).

Also actively involved in the harmonization of environmental data in research and monitoring is the International Council of Scientific Unions (ICSU), represented by its Scientific Committee for the Problems of the Environment (SCOPE) and its Committee on Data for Science and Technology (CODATA) [BDC+95]. The Norwegian SAMPO project uses ESRI's ArcView to catalog its spatial data holdings [MR95]. The Austrian Ministry of the Environment has developed the Central European Environmental Data Request Facility (CEDAR) [PK92], which can be reached at http://www.cedar.univie.ac.at. Other efforts include the CIMI system of the Dutch Ministry of Transportation, Public Works and Water Management [Kug95], the Australian FINDAR system [JSTC91] and the New South Wales Department of Conservation and Land Management's Data Directory [MF94, MB94].

Coordination between this great variety of efforts is difficult. As we will see in Sect. 5.3, the newly founded European Environment Agency will have an important role to play here. One promising effort concerns the development of a common European geodata standard. With strong support from the European Center of Normalization, Germany and Belgium's Geographic Data Files (GDF) are generally considered the frontrunner [Ost95]. Further standardization is required, however. Environmental phenomena do not stop at national borders. In this domain, international cooperation on a broad scale is essential for making progress.

## 5.2 Metadata in the U.S. National Information Infrastructure

Since the early 1980s, the U.S. Government has been working intensively on creating a National Spatial Data Infrastructure (NSDI). A major motivation for this effort was to abolish the notorious incompatibilities among the internal formats used by various government agencies. Examples include DLG, TIGER/Line, and GRASS of the U.S. Geological Survey, DIGEST and the Vector Product Format (VPF) of the Defense Mapping Agency (DMA), and DX90 of the National Ocean Service. The parallel use of such a variety of standards led to considerable expenses to the taxpayer that could at least in part have been avoided.

Most of the early efforts on NSDI were coordinated by the U.S. Geological Survey, an agency under the supervision of the U.S. Department of the Interior. One of the first major results was the development of the Spatial Data Transfer Standard (SDTS), a Federal Information Processing Standard to facilitate the online exchange of spatial data [Uni92]. The goal is to accommodate different spatial data models, to preserve topologies, and to maintain even complex relationships, as data is transferred across different computer platforms and software systems. Other than many existing standards (such as VPF), the SDTS is not an exchange format. It rather provides guidelines that need to be translated into a native application-specific format before they can be used. Most GIS vendors provide interfaces and tools for that purpose [ESR95c].

In the early 1990s, the NSDI was integrated into the National Information Infrastructure (NII). NII is a program coordinated by the White House to reorganize and renew the computer infrastructure throughout all levels of government. A middle-term goal of NII is to offer e-mail and Web access to every single government employee. Some of the key technical cornerstones of NII are:

- client/server technology
- TCP/IP-based intranets
- Internet connectivity
- relational databases.

NSDI has since been coordinated by a working group called the Federal Geographic Data Committee (FGDC), which is composed of representatives of the Departments of Agriculture, Commerce, Defense, Energy, Housing and Urban Development, the Interior, State, and Transportation; the Environmental Protection Agency; the Federal Emergency Management Agency; the Library of Congress; the National Aeronautics and Space Administration; the National Archives and Records Administration; and the Tennessee Valley Authority. The committee is chaired by the Department of the Interior, represented by the U.S. Geological Survey.

In May 1994, the FGDC published a draft for the new Content Standards for Digital Geospatial Metadata [Uni94a], which was later approved by the National Institute of Standards and Technology as a Federal Information Processing Standard. The implementation of the standard is based on the Executive Order 12906, "Coordinating Geographic Data Acquisition and Access: The National Spatial Data Infrastructure," which was signed on April 11, 1994, by President Clinton [Uni94b]. In addition to providing a long-needed political foundation for the NSDI, the order requires all government agencies to use the FGDC Content Standards for documenting all new geospatial data it collects or produces as of April 11, 1995.

While both the SDTS and the FGDC Content Standards refer to metadata about spatial data, they have distinctly separate functions. The SDTS is a language for communicating spatial data across different platforms without losing any structural or topological information. The FGDC Content Standards, on the other hand, specify the kind of annotative metadata that federal agencies are required to collect on a spatial data set they maintain. The only two sections that both standards have in common concern data quality and the data dictionary information; we will discuss this in detail later on.

### 5.2.1 The Government Information Locator Service

The objective of the Government Information Locator Service (GILS) was to build a national metainformation system to serve as a navigation and organization aid. GILS supports users for searching public databases and accessing official documents and files. Moreover, it is used as an internal documentation and organization tool.

GILS is organized as a decentralized cooperation of existing domain-specific information systems in ministries and subordinate government agencies. The integration of these systems is performed *ex post* and *bottom-up*, i.e., without compromising their autonomy. All federal agencies, however, are obliged to document new data sets in accordance with GILS guidelines, and gradually to provide documentation for older data sets as well. GILS is coordinated by the Office of Management and Budget (OMB), the National Archive and Record Association (NARA), and the National Institute of Standards (NIST). In summary, GILS serves to

- *document* public information resources held by government agencies;
- *describe* the information available;
- help users to query and retrieve the information *(navigation aid)*.

The GILS metadata is organized in GILS Core Records, which have the following fields (or attributes):

1. *Title*: a concise, non-formatted textual description of the data set
2. *Abstract*: a description of the data set in free text, typically a few lines long
3. *Acronym*
4. *Agency program*: primary usage of the data set at the responsible agency
5. *Controlled vocabulary*: keywords chosen from a thesaurus or another limited vocabulary
6. *Local subject term*
7. *Originator*: contact person.

GILS Core Records for *geographic* data sets obey the FGDC Content Standards for Digital Spatial Metadata (see Sect. 5.2.3).

Figures 5.2–5.3 illustrate the GILS record structure with examples from the Web site of the Environmental Protection Agency (EPA). The EPA represents one of the most active GILS user communities. EPA uses GILS both for supporting access to its data sets by citizens, i.e., as a navigation aid, and for its internal information management.

Compared to other metainformation systems, such as CDS (Sect. 5.3) or UDK (Sect. 5.4), the level of detail of GILS Core Records is quite coarse. This is not surprising, given the GILS objective to be not only a navigation aid but also an attempt to organize government-held information in a single framework. In order to motivate a large number of government agencies, the entry level for participation has been kept low. It is particularly important to minimize the time effort to produce a GILS Core Record; ideally, it should not take more than a few minutes.

User interfaces to GILS vary between agencies although they are generally WWW-based. Key functionalities include fulltext search and browsing, possibly supported by a thesaurus. Figures 5.3–5.5 illustrate these functionalities on the EPA's GILS Web pages. Web browsers are also used to support

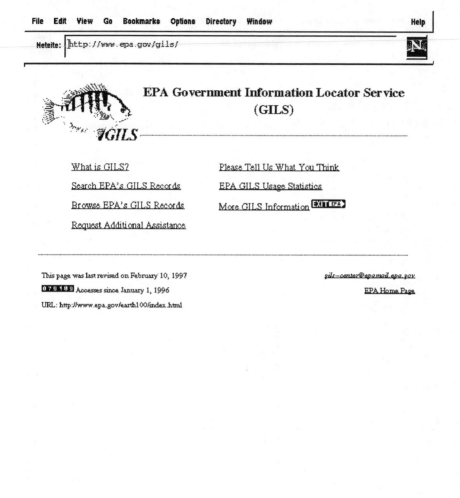

**Fig. 5.2.** Entry page of EPA's GILS Web site

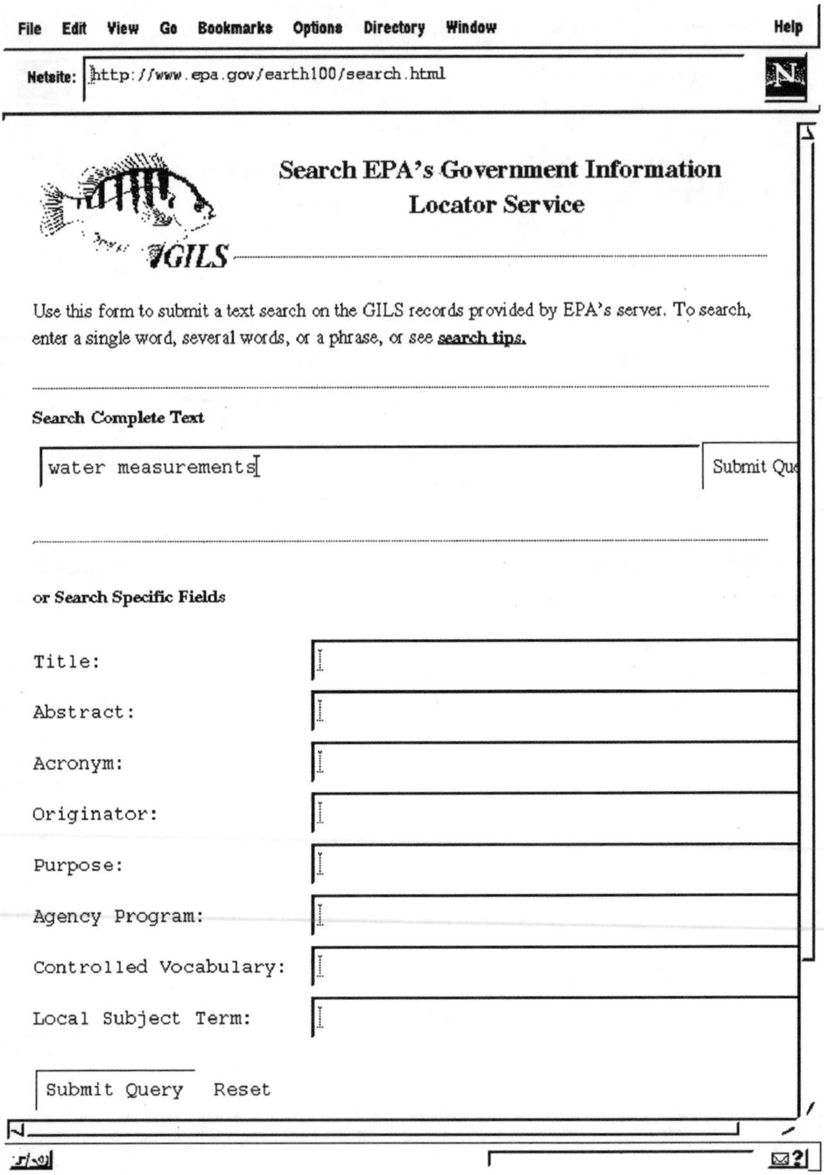

| File   Edit   View   Go   Bookmarks   Options   Directory   Window | Help |

Netsite: http://www.epa.gov/earth100/search.html

## Search EPA's Government Information Locator Service

*GILS*

Use this form to submit a text search on the GILS records provided by EPA's server. To search, enter a single word, several words, or a phrase, or see **search tips.**

**Search Complete Text**

water measurements                                                    Submit Que

**or Search Specific Fields**

Title:

Abstract:

Acronym:

Originator:

Purpose:

Agency Program:

Controlled Vocabulary:

Local Subject Term:

Submit Query   Reset

**Fig. 5.3.** Searching the EPA GILS records

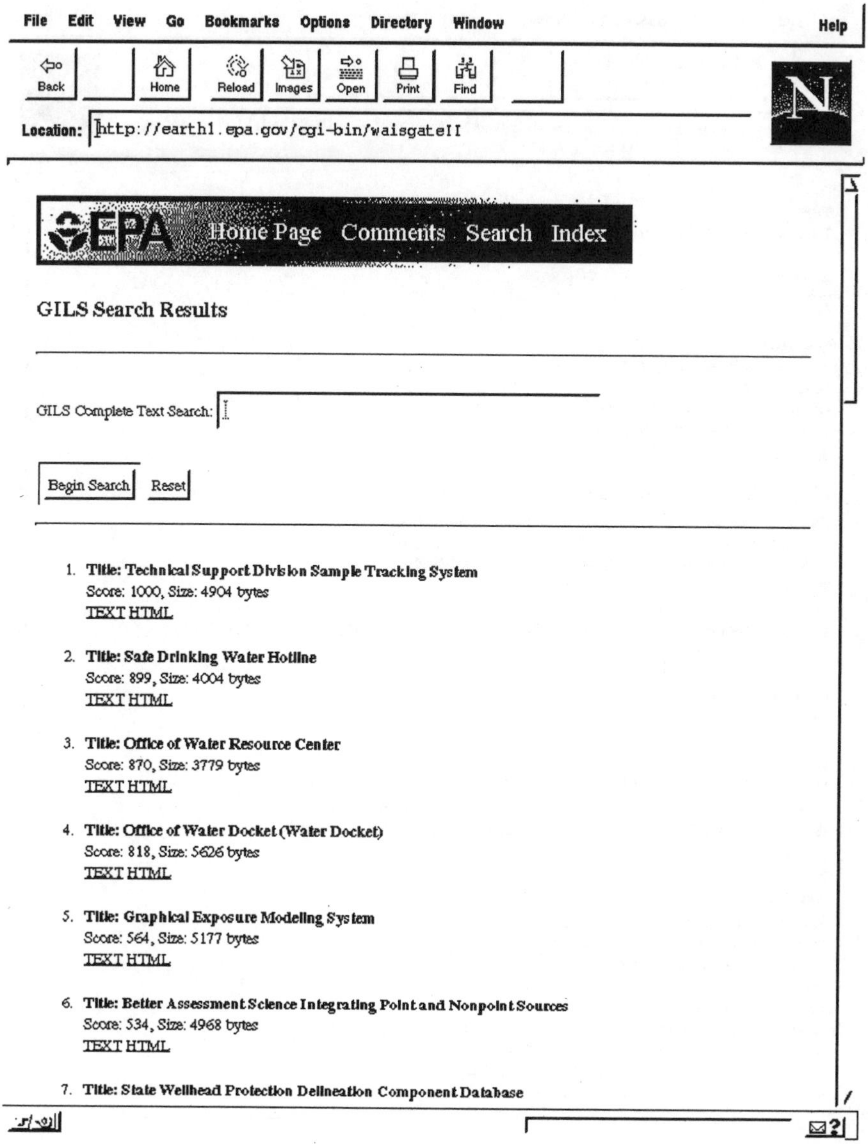

File   Edit   View   Go   Bookmarks   Options   Directory   Window                                    Help

⇦o        🏠         🔄        📑        ⇨o        🖶        🔍
Back      Home     Reload    Images    Open      Print     Find

Location: http://earth1.epa.gov/cgi-bin/waisgateII

**EPA**   Home Page   Comments   Search   Index

**GILS Search Results**

GILS Complete Text Search: |

Begin Search   Reset

1. **Title: Technical Support Division Sample Tracking System**
   Score: 1000, Size: 4904 bytes
   TEXT HTML

2. **Title: Safe Drinking Water Hotline**
   Score: 899, Size: 4004 bytes
   TEXT HTML

3. **Title: Office of Water Resource Center**
   Score: 870, Size: 3779 bytes
   TEXT HTML

4. **Title: Office of Water Docket (Water Docket)**
   Score: 818, Size: 5626 bytes
   TEXT HTML

5. **Title: Graphical Exposure Modeling System**
   Score: 564, Size: 5177 bytes
   TEXT HTML

6. **Title: Better Assessment Science Integrating Point and Nonpoint Sources**
   Score: 534, Size: 4968 bytes
   TEXT HTML

7. **Title: State Wellhead Protection Delineation Component Database**

**Fig. 5.4.** EPA GILS search results for the term *water measurements*

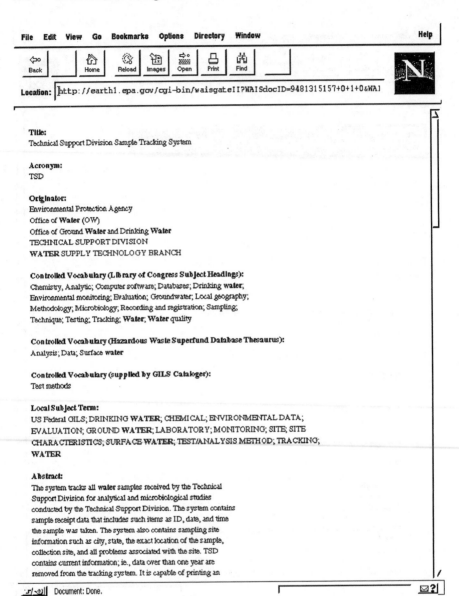

**Fig. 5.5.** EPA GILS catalog entry

the entry and validation of GILS Core Records. The fact that different agencies have different interfaces is drawing mixed reviews. On the one hand, it allows agencies to emphasize their specific agenda and the relevant data sets. On the other hand, users who work with GILS at more than one agency are generally less than enthusiastic about the lack of a common look and feel.

GILS uses the Z.39.50 protocol to access several distributed servers simultaneously. There exists a WWW Z.39.50 conversion routine to translate GILS queries that have been posed using a Web browser into a format compatible with Z.39.50. Conversely, the results of a query are reconverted to be displayed in the chosen Web browser.

### 5.2.2 The Spatial Data Transfer Standard

The Spatial Data Transfer Standard (SDTS) [Uni92] was designed to facilitate the online transfer of the full range of geographic and cartographic data. Both vector and raster data for a large variety of data models can be exchanged across heterogenous hardware and software platforms using SDTS. The standard is structured into three main parts; the subsequent presentation follows the overview of Fegeas et al. [FCL92].

**Part 1: Logical Specification.** This part contains the logical specification of the entities and data objects used to describe different GIS data models. It consists of three major sections and provides guidelines on how spatial and nonspatial objects (simple or composite) are to be organized, named, and structured.

The first section presents a conceptual model of spatial data. It describes the real world as a set of *entities* (cities, rivers, factories, etc.), each characterized by attributes, which are assigned attribute values. The model then goes on to define zero-, one-, and two-dimensional *spatial objects* (such as points, lines, and polygons) and the relationships between entities and spatial objects.

The reader should take note of the particular use of the term *entity*, which in this standard has been chosen to describe a real-world phenomenon, whereas the term *object* is reserved for the digital representation of an entity. With reference to the terminology used in this book (cf. Chap. 1), SDTS entitities correspond to environmental objects and SDTS objects to environmental data objects.

In analogy to the standard entity-relationship literature, SDTS uses the term *entity type* to describe a set of similar entities; in that context the single entities are also called *entity instances*. The term *feature*, finally, which is still very common in the geoscientific community, is here defined as both a real-world entity and its object representation, it thus encompasses the two SDTS terms *entity* and *object*.

The second section of Part 1 is devoted to data quality. It specifies five portions of a data quality report: lineage, positional accuracy, attribute accuracy, logical consistency, and completeness. The lineage portion describes source and update material (with dates), methods of derivation, transformations, and other processing history. Positional accuracy is concerned with how closely the locational data represent true locations. Attribute accuracy is similarly concerned with non-locational descriptive data. Logical consistency refers to the fidelity of encoded relationships in the structure of the spatial data (e.g., the degree to which topological relationships have been verified). The completeness portion includes information about geographic area and subject matter coverage. Note that large parts of this second section of Part 1 are replicated in feature group 2 (data quality information) of the FGDC Content Standards.

The third section of Part 1 constitutes the largest portion of the whole standard; it specifies detailed logical transfer format constructs and specifications for SDTS transfer data sets. An SDTS transfer is organized into modules with records, fields, and subfields. Thirty-four module types are specified as detailed field and subfield record layout specification tables, designed to include many kinds of information: global, data quality, feature and attribute data dictionary, coordinate reference, spatial object, and associated attribute and graphic symbology information. The data dictionary portion, which conveys the meaning and structure of entity and attribute data, is divided into three module types: definition, domain, and schema. Parts of the data dictionary portion are replicated in feature group 4 (entity and attribute information) of the FGDC Content Standards.

**Part 2: Data Content Registry.** This part provides data content standards by specifying a model for the definition of spatial entity types, attributes, and attribute values. The underlying idea of this part of the standard is that there is a need for common definitions of spatial features (resp. entities). In that sense, this part is nothing but a thesaurus. It contains a list of about 200 topographic and hydrographic entity types with 244 attributes, plus a list of about 1200 terms that are in a synonym or subtype relationship to any of those standard or primary terms. It is foreseen by the designers of the standard that this section will be subject to continual updates and extensions.

**Part 3: Physical Structure.** This part specifies the implementation of the transfer using the ISO 8211 international standard for information interchange. The ISO standard itself is embedded in the SDTS to ensure that data can be transferred to any computing environment. The U.S. Geological Survey has developed a public domain software function library to assist in encoding and decoding SDTS data into ISO 8211 format.

It is important to keep in mind that the SDTS and ISO 8211 are separate standards. ISO 8211 is an international data exchange format that can be used to transfer any type of data, not just spatial data. ISO 8211 provides a means of transferring data records and their description across heterogenous hardware and software platforms. It requires, however, that the *content* and the *meaning* of the data records are defined by the user. In that sense, the SDTS can be considered a user of ISO 8211.

The SDTS is designed such that Parts 1 and 2 are independent of Part 3, which is specific to ISO 8211. If necessary, the SDTS could replace Part 3 by another version that uses a different implementation format without affecting Parts 1 and 2. ISO 8211 was chosen so that the SDTS could use an existing general-purpose transfer standard rather than having to develop a new SDTS-specific format. It is designed to work for any media, including communication lines. ISO 8211 is self-describing. An ISO 8211 file (called a *Data Descriptive File (DDF)*) contains both data and the description of the data. The *Data Descriptive Record (DDR)* is fixed; it contains the structure and description of the data. The *Data Records (DRs)* are of variable size; they contain the actual data. There is always one DDR in a DDF, and one or more DRs.

Given the great complexity of the standard, the designers also introduced a concept called the *profile*, which is a kind of customization of the standard for a particular data model. If a new data model is to be supported, the interested parties may specify those options of the standard that are needed to support that data model. This subset of options can then be submitted for approval as its own Federal Information Processing Standard (FIPS) and, once approved, is added to the SDTS as a new SDTS profile.

Currently, there exists a Topological Vector Profile (TVP) for vector data with full and explicit topology. Another profile that is about to be approved is a raster profile for image and gridded data. Under consideration are further vector profiles for network/transportation data, for nontopological nautical chart and hydrographic data, and for CAD data.

### 5.2.3 The FGDC Content Standards

The FGDC Content Standards for Digital Geospatial Metadata define metadata as *data about the content, quality, condition, and other characteristics of data.* They structure the spatial metadata into the following seven groups of features. Only the first (identification information) and the last feature group (metadata reference information) are obligatory; the remaining ones are optional.

**Group 1: Identification Information.** This feature group contains the basic metainformation about a given data set. It is structured into the following attributes:

- Textual description.
- Information about the time period described.
- Spatial reference: A minimum bounding rectangle is required. Optionally, one can provide a more detailed polygonal description.
- Keywords: They can be freely chosen, but need to be associated with a term from the relevant thesaurus. One keyword about the theme of the

data set is obligatory. Optionally, one can provide further keywords that refer to the theme, the space, or the time corresponding to the data set in question.

- Person or organization to contact for more information about the data set (optional).
- Access constraints and security information (optional).
- Information about the technical representation of the data set: special software, operating system, file name, data set size (optional).

**Group 2: Data Quality Information.** This feature group contains general information about the quality of the data set. In addition to an assessment of the accuracy and consistency of the data, this includes metadata about the data source (*lineage*) and about completeness. Note that this feature group replicates the content (but not the structure) of the SDTS's data quality section (Part 1, Sect. 2).

**Group 3: Spatial Data Organization Information.** This feature group contains information on which mechanism was used to represent spatial information in the data set. At present, the standard supports a generic mechanism to represent raster data, and SDTS and VPF to represent vector data. The SDTS section is based on Part 1 of the SDTS specification. The fact that both SDTS and VPF were included explicitly shows how the designers of the standard sometimes had to sacrifice conciseness and clarity in order to obtain approval from all participants. It was not possible to move all government agencies towards a single standard for representing vector data. Among other reasons, this is due to large amounts of essential legacy data, whose conversion would exceed the available resources of the respective agencies.

**Group 4: Spatial Reference Information.** This feature describes the projection and coordinate system used (e.g., Mercator or Miller_Cylindrical).

**Group 5: Entity and Attribute Information.** This feature group allows the user to describe the information content of the data set using the entity-relationship model. The SDTS's data dictionary information is captured in this feature group. There is common agreement that this section of the standard is too superficial and should be redesigned in future versions of the standard.

**Group 6: Distribution Information.** This feature group contains information about the distributor of the data set and about options for obtaining it. The distributor usually corresponds to the contact person/organization listed in the identification information (feature group 1). The order information includes data about the possible modes of communication (modem, e-mail, etc.) and about the transfer formats used (e.g., the ARC/INFO Export format, the Initial Graphics Exchange Standard (IGES), or ASCII).

**Group 7: Metadata Reference Information.** This obligatory feature group serves for storing what could be called *meta-metadata*. This includes information about the last update of the metadata, the latest and the next review of the metadata, the party responsible for the metadata, and access and security constraints.

In summary, the FGDC Content Standards represent an impressive effort to establish a uniform way to document digital geospatial data sets. While mainly targeted at the description of geographic data, it also provides a solid basis for an environmental metadata system. Such an extension would entail a more detailed semantic framework, especially with regard to theme-related information.

## 5.3 The Catalogue of Data Sources

The European Union (EU) has been working on similar issues, especially since the 1994 foundation of its European Environment Agency (EEA), located in Copenhagen. In comparison to the American activities, the EEA efforts to build a Catalogue of Data Sources have a wider focus, concentrating not only on spatial data, but on environmental data in a more general sense. On the other hand, the results obtained so far are not quite as concrete as the FGDC recommendations described above.

The original goal of the EU activities was the implementation of an integrated European environmental information system. Based on the results of a previous project called CORINE CDS (1985–1989), in 1992 the EU commissioned a study entitled "Catalogue of Data Sources for the Environment – Analysis and Suggestions for a Meta-Data System and Service for The European Environment Agency" [Eur93]. It became clear immediately that the construction of a European environmental information system from scratch is neither economically feasible nor politically viable. Many member countries already have some kind of national environmental information system. A European system should take advantage of these developments and attempt a bottom-up integration of the systems that are already functional. Devised as a *meta*-information system, such a CDS would only store descriptions of data sets that are locally available.

The study recommends the simultaneous realization of the following two architectures:

– a standalone variant that is updated periodically based on current information from the member countries;
– a networked variant, which has online connections with a variety of national catalogs and is only usable in connection with those.

Since the study was written (1992/93), the percentage of computers that are networked, usually including some connection to the Internet, has grown considerably. The first architecture option therefore seems increasingly obsolete. In turn, it should be ensured that the central catalog provides some base functionalities independently of the current state of the national catalogs and the connections to them. This can easily be achieved by making local copies of a subset of the metadata periodically. For distribution and update purposes, the study recommends the usage of CD-ROMs. Once again, however, this recommendation has been superseded by the rise of the Internet. With the possible exception of a few isolated sites, anybody using a system like CDS will have Internet access by the time CDS will have become widely available. Other recommendations include the storage of the data in a relational database system using text fields, and the integration of a multilingual thesaurus.

The study does not propose a concrete format for the metadata, comparable to the detailed specifications of the FGDC or the UDK (see Sect. 5.4). The authors suggest instead forming some synthesis of the existing proposals of the member countries and of the United States. Moreover, they propose a simple class model. The three main classes, together with some of their attributes, are:

1. *Institutions*: name, description, functions, language
2. *Activities/Projects*: name, description
3. *Products*: name, description, contents

The remaining five classes serve to represent secondary information about the objects in classes 1, 2, and 3:

4. *Addresses*: address, geolocation
5. *Stations*: name, scope, description, equipment
6. *Communication*: telephone/fax/telex, networks
7. *People/Persons*: name, title, functions, language
8. *Data Sets*: name, object/parameter, unit, scale

Regardless of the particular class structure chosen, it seems somewhat questionable whether a single-layer taxonomy like the one proposed would ever be able to capture the extreme heterogeneity that resulted from a synthesis of the environmental data and metadata schemes throughout Europe. On the one hand, there will always be entity types that do not fit into the given scheme. On the other hand, there has to be a formal mechanism to refine a given entity class in order to serve the local requirements of a particular agency in an optimal manner. A multi-layer taxonomy, i.e., a class hierarchy with an inheritance mechanism, seems to be better suited for this purpose (see also the discussion of the UDK, Sect. 5.4.2).

In the meantime, the European Environment Agency has intensified their related efforts and started a European Topic Centre on Catalogue of Data Sources (ETC/CDS), based in Hanover (Germany). In 1995, ETC/CDS commissioned a first CDS prototype. The design of this prototype deviates from the 1993 study in several important aspects [Eur96, Eur97e, Eur98].

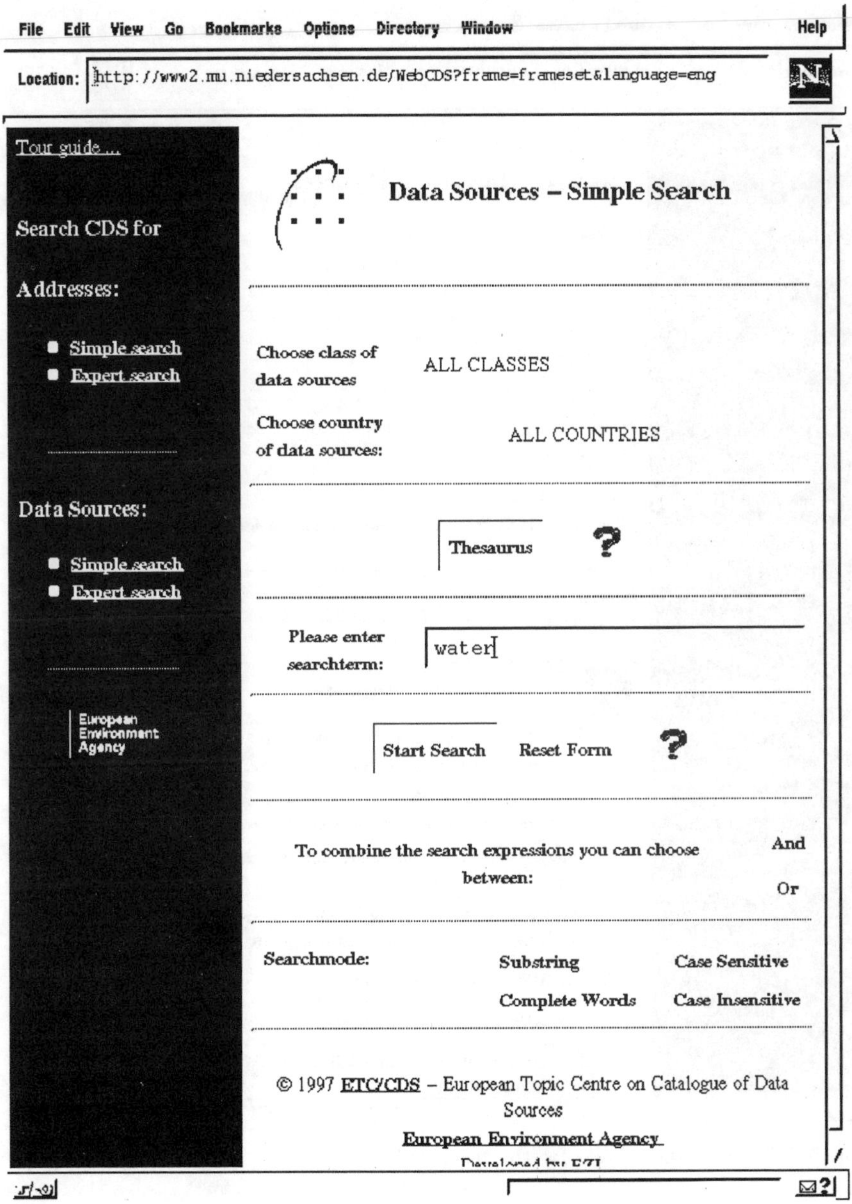

| File | Edit | View | Go | Bookmarks | Options | Directory | Window | Help |

Location: http://www2.mu.niedersachsen.de/WebCDS?frame=frameset&language=eng

Tour guide....

Search CDS for

Addresses:

- Simple search
- Expert search

Data Sources:

- Simple search
- Expert search

European
Environment
Agency

**Data Sources – Simple Search**

Choose class of
data sources        ALL CLASSES

Choose country
of data sources:        ALL COUNTRIES

Thesaurus    ?

Please enter
searchterm:    water

Start Search    Reset Form    ?

To combine the search expressions you can choose    And
between:    Or

Searchmode:    Substring    Case Sensitive
    Complete Words    Case Insensitive

© 1997 ETC/CDS – European Topic Centre on Catalogue of Data
Sources
European Environment Agency
Developed by ETI

**Fig. 5.6.** Entry page of the Catalogue of Data Sources

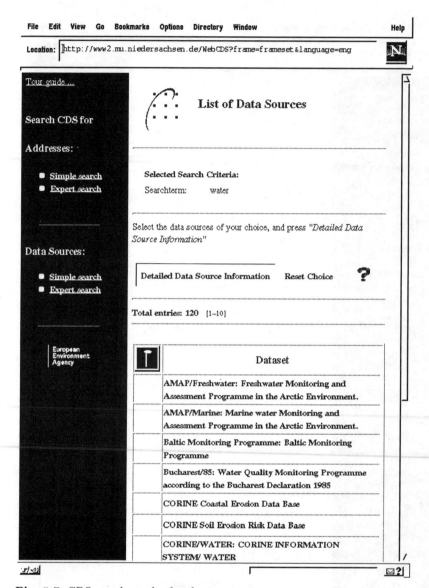

Fig. 5.7. CDS search results for the term *water*

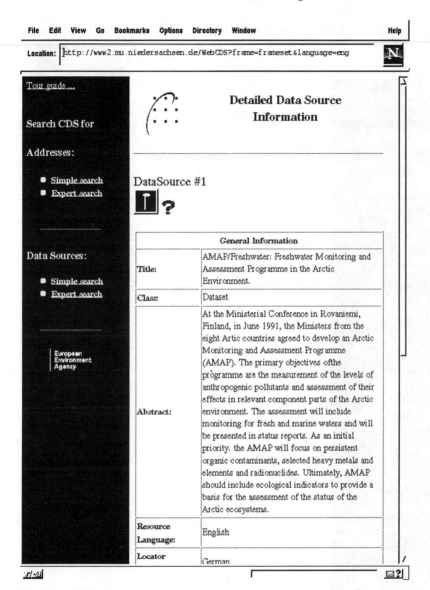

**Fig. 5.8.** CDS catalog entry

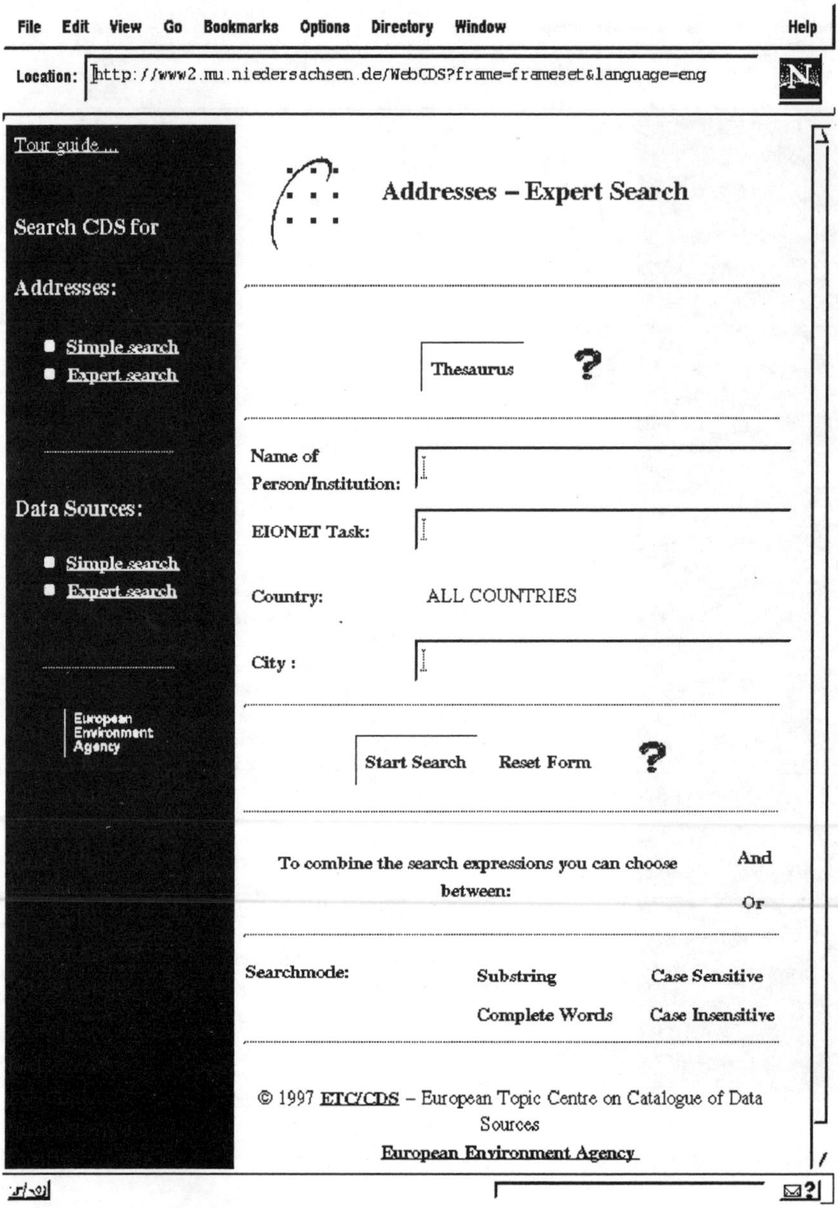

Fig. 5.9. CDS search for an address

On the one hand, CDS is now focusing on the *locator* aspect of a metain-formation system. The objective is to provide answers to the following questions:

- Who in Europe holds environmental information on a given topic?
- In which format is it stored?
- How can one access it?

An important goal of CDS is to support the operation of EIONET, the EEA's Environmental Information and Observation Network.

On the other hand, the CDS data model has been redesigned. While not object-oriented in the strict sense, the new data model still integrates several important aspects of object-orientation. Objects in CDS are called *CDS records*. Following a recent draft document [Eur96], CDS records are grouped into several classes (here called *categories*), depending on the kind of data source they describe:

1. *Environmental Subjects* with subclasses
   1.1. *Activities*
   1.2. *Data Sets*
2. *Institutions*
3. *People*

The class *Environmental Subjects* is an abstract class because each environmental subject is either an *activity* (e.g., a monitoring program) or a *data set* (e.g., a measuring series). *Institutions* and *people* are responsible for certain environmental subjects. Although one environmental subject may be associated with several institutions and people, one of these is typically designated as the main contact.

CDS records exist at three levels of detail, thus creating an inheritance hierarchy below the classes listed above:

- The *core level* comprises those attributes that are considered to be of basic importance. They are recommended to be used in all national CDS implementations to maintain compatibility.
- *Level 1* contains core level attributes and additional attributes needed by the EEA. The additional attributes serve in particular to provide detailed information on the spatial, temporal, and technical characteristics of a data source.
- *Level 2* contains core level attributes and additional attributes to be defined by national authorities for their local CDS implementations.

The current CDS architecture consists of three subsystems: a Windows application called *WinCDS*, a Web interface called *WebCDS*, and a multilingual thesaurus called *GEMET*. We discuss these three components in turn.

*WinCDS* is an MS Access application designed to support the *registering* of data sources with CDS. It allows information suppliers to describe their data holdings in accordance with the CDS requirements. The software is available on CD-ROM. It may also be downloaded from the CDS Web server (http://www.mu.niedersachsen.de/cds).

*WebCDS* is a Java-based tool for *finding* environmental information via CDS. WebCDS supports keyword searches for *addresses* and *data sources*. More complex search operations are possible as well. An interesting feature of the present WebCDS implementation is its two-window concept: one window serves to control the search, the other one is for displaying the results. Figures 5.6–5.9 show several pages from a WebCDS prototype.

*GEMET (General European Multilingual Environment Thesaurus)* is a multilingual thesaurus designed to support searches in CDS. It is implemented in British English with equivalents in American English. Equivalents in Danish, Dutch, French, German, Italian, Norwegian, and Spanish are partially available. GEMET has been compiled on the basis of the following national thesauri:

- EnVoc Thesaurus of the United Nations Environmental Program (UNEP)
- EuroVoc Thesaurus of the European Parliament
- Lexique Environnement – Planète of the French Ministry of the Environment
- Multilingual Environment Thesaurus (MET) of the Netherlands Current Research Information Agency (NBOI)
- Thesaurus de Medio Ambiente of the Spanish Ministry of Public Works, Transport, and Environment (MOPTMA)
- Thesaurus Italiano per l'Ambiente (TIA) of the Italian National Research Council (CNR)
- Umweltthesaurus of the German Federal Environmental Agency (UBA)

The compilation yielded close to 5 000 descriptors, which were arranged in a hierarchical and thematical classification scheme made up of 34 groups. Polyhierarchy was kept to a minimum.

## 5.4 UDK: Environmental Data Catalog

The *UDK (Umwelt-Datenkatalog = Environmental Data Catalog)* is a metainformation system and navigation tool that documents collections of environmental data from the government and other sources. These data sets may be available either online or by request to the responsible data administrator. Potential users of the system include government agencies, industry, and the general public. Similar to CDS, the UDK helps them to obtain answers to the following questions:

- Which relevant information is in principle available for a given problem?
- Where is this information stored?
- How can this information be retrieved?

The UDK design presented in this section is the result of several years of research and development [Les89, LS94]. In 1990, the Environmental Ministry of the State of Lower Saxony launched a research project with funding from the German Federal Environmental Agency. Two years later, an international working group was formed to oversee the UDK design and its further development into a practical software tool. In 1994, Austria passed an Environmental Information Law that introduced the UDK as the official navigation tool for all environmental information on record. In 1995, first versions of the UDK were made available in Austria and the German states of Baden-Württemberg and Lower Saxony. Other German states and several other European countries are considering to adapt the UDK as well. Several key concepts of the UDK have found their way into other systems, including the Catalogue of Data Sources described above.

## 5.4.1 Object Model

The UDK is based on a three-way object model that is very similar to the data model described in the introduction (Fig. 1.1). The UDK distinguishes between environmental objects, environmental data objects, and UDK (meta-)objects. Each real-world environmental object is described by a collection of environmental data objects. Each environmental data object is in turn associated with exactly one *UDK object*, i.e., a metadata object that specifies its format and contents.

On the screen, each such UDK object can be represented by one or more screen layouts; see Fig. 5.10 for examples from an earlier UDK version. The first screen layout contains some administrative information (object name, object ID, and keywords), a text description, and the address of the agency responsible for the maintenance of this UDK object and the underlying environmental data object. The second screen layout contains some more technical information about the environmental data object. This includes detailed data about the information content, the capturing method and its accuracy, the spatial extent, and the validity of the object. Spatial information can be specified using either coordinates or (as in this example) denominations of administrative entities.

UDK objects may exist for environmental data objects at various aggregation levels simultaneously. Consider, for example, a national groundwater database that contains a large number of measurements from all over the country. There is one UDK object representing this database as a whole. In addition there may be one UDK object each for the measurements from a certain county, there may be UDK objects representing the measurements from a particular station, and there may even be UDK objects that represent single measurements. There may also be UDK objects for groupings that are orthogonal to this primary aggregation hierarchy, such as UDK objects representing the measurements that were taken in a given month.

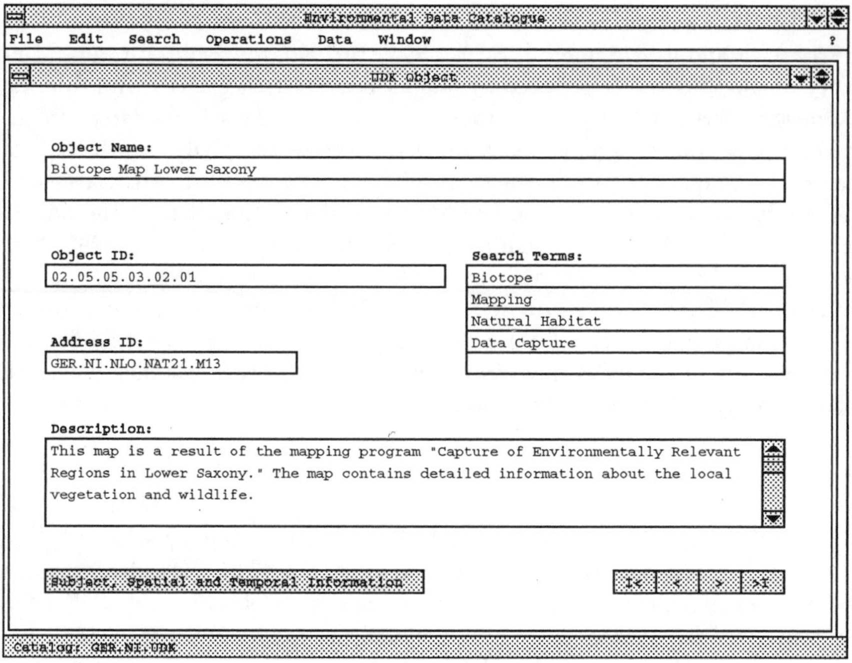

```
┌─────────────────────────────────────────────────────────────────────┐
│▤            Environmental Data Catalogue              ▼│▲│
│ File   Edit   Search   Operations   Data   Window                   ?│
│ ┌─────────────────────────────────────────────────────────────┐▼│▲│
│ │▤                  UDK object                            │
│ │                                                              │
│ │    Object Name:                                              │
│ │    ┌────────────────────────────────────────────────┐       │
│ │    │Biotope Map Lower Saxony              ·          │       │
│ │    └────────────────────────────────────────────────┘       │
│ │                                                              │
│ │    Object ID:                       Search Terms:            │
│ │    ┌───────────────────────┐        ┌──────────────────┐    │
│ │    │02.05.05.03.02.01      │        │Biotope           │    │
│ │    └───────────────────────┘        ├──────────────────┤    │
│ │                                     │Mapping           │    │
│ │                                     ├──────────────────┤    │
│ │                                     │Natural Habitat   │    │
│ │    Address ID:                      ├──────────────────┤    │
│ │    ┌───────────────────────┐        │Data Capture      │    │
│ │    │GER.NI.NLO.NAT21.M13   │        └──────────────────┘    │
│ │    └───────────────────────┘                                │
│ │                                                              │
│ │    Description:                                              │
│ │    ┌──────────────────────────────────────────────────┐▲│  │
│ │    │This map is a result of the mapping program "Capture of Environmentally Relevant │
│ │    │Regions in Lower Saxony." The map contains detailed information about the local  │
│ │    │vegetation and wildlife.                          │  │  │
│ │    └──────────────────────────────────────────────────┘▼│  │
│ │                                                              │
│ │  ┌──────────────────────────────────────────┐   ┌─┬─┬─┬─┐   │
│ │  │Subject, Spatial and Temporal Information  │   │I<│<│>│>I│ │
│ │  └──────────────────────────────────────────┘   └─┴─┴─┴─┘   │
│ └──────────────────────────────────────────────────────────────┘
│ Catalog: GER.NI.UDK                                                  │
└─────────────────────────────────────────────────────────────────────┘
```

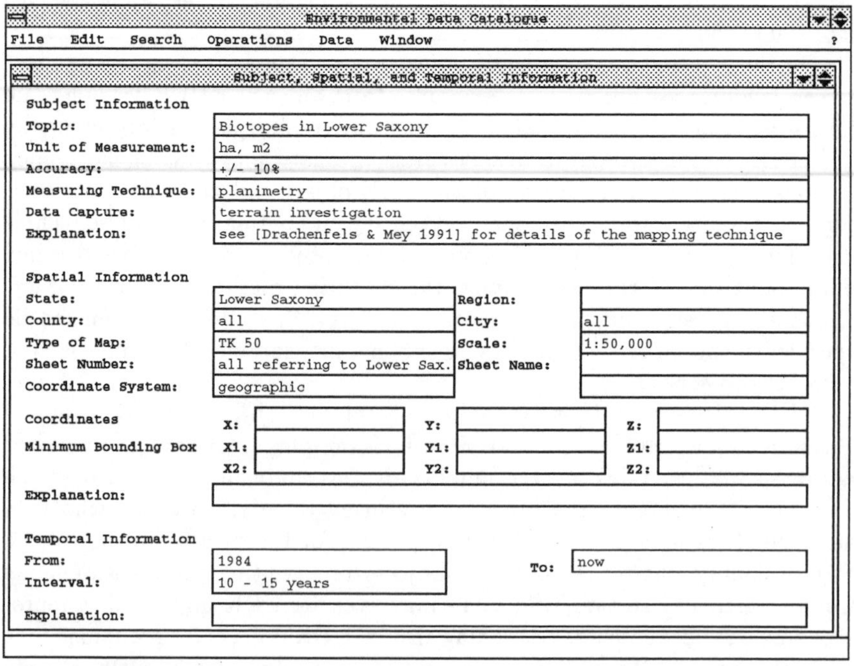

```
┌─────────────────────────────────────────────────────────────────────┐
│▤            Environmental Data Catalogue              ▼│▲│
│ File   Edit   Search   Operations   Data   Window                   ?│
│ ┌─────────────────────────────────────────────────────────────┐▼│▲│
│ │▤        Subject, Spatial, and Temporal Information       │
│ │ Subject Information                                          │
│ │ Topic:              ┌──────────────────────────────────────┐│
│ │                     │Biotopes in Lower Saxony              ││
│ │ Unit of Measurement:│ha, m2                                ││
│ │ Accuracy:           │+/- 10%                               ││
│ │ Measuring Technique:│planimetry                            ││
│ │ Data Capture:       │terrain investigation                 ││
│ │ Explanation:        │see [Drachenfels & Mey 1991] for details of the mapping technique│
│ │                                                              │
│ │ Spatial Information                                          │
│ │ State:      ┌───────────────┐    Region: ┌─────────────┐    │
│ │             │Lower Saxony   │            │             │    │
│ │ County:     │all            │    City:   │all          │    │
│ │ Type of Map:│TK 50          │    Scale:  │1:50,000     │    │
│ │ Sheet Number:│all referring to Lower Sax.│Sheet Name:  │    │
│ │ Coordinate System:│geographic │            │             │    │
│ │                                                              │
│ │ Coordinates          X:┌────┐    Y:┌────┐    Z:┌────┐        │
│ │ Minimum Bounding Box X1:│    │   Y1:│    │   Z1:│    │       │
│ │                      X2:│    │   Y2:│    │   Z2:│    │       │
│ │ Explanation:         ┌──────────────────────────────────┐   │
│ │                                                              │
│ │ Temporal Information                                         │
│ │ From:       ┌───────────────┐          To: ┌──────────────┐ │
│ │             │1984           │              │now           │ │
│ │ Interval:   │10 - 15 years  │                              │ │
│ │ Explanation:┌──────────────────────────────────────────┐   │
│ └──────────────────────────────────────────────────────────────┘
└─────────────────────────────────────────────────────────────────────┘
```

Fig. 5.10. Two screen layouts representing a UDK object

There are two reasons for this great flexibility in defining UDK objects at various levels of aggregation. First, powerful aggregation facilities are crucial for improving the usability and acceptance of a system like the UDK. Empirical studies have shown that the overwhelming number of queries in such a context refer to aggregated data rather than detailed source data. For example, citizens may be concerned about the ozone concentration in their neighborhood on a certain day; it is rather unlikely that they would want to know the exact concentration at a certain measuring station at an exact time. Second, aggregation semantics differ greatly between different user communities. Some people may have to aggregate over time, others over space, and yet others by topic. To appeal to a large user community, the UDK system must be able to accommodate those different needs.

Although it is therefore desirable to handle the creation (and deletion) of UDK objects with great flexibility, the decision to create a new object has to be based on a cost/benefit analysis, depending on the particular applications a user has in mind. The effort to create and maintain a UDK object is not negligible. Recent empirical data suggests that creation takes close to one person-day on average. Maintenance involves not only the occasional update of attributes but also the dynamic tracking of semantic associations between UDK objects and the corresponding environmental data objects; see Sect. 5.4.3 for further details. At this time, most of the related work is performed by specialized personnel from higher-level government agencies or consulting firms, and is therefore relatively expensive. It is unlikely that the work can be delegated to less-qualified support staff in the near future. The idea of leaving the creation of UDK objects to local domain experts (biologists, chemists, etc.) is also unrealistic at the present time. The process is still too technical and time-consuming for someone who is not a UDK expert.

In earlier versions of the UDK, UDK objects were identified by their position in the *primary tree*, a directed graph whose nodes correspond to the UDK objects and whose edges represent responsibilities of agencies and departments for particular sets of UDK objects, as well as *part-of* relationships between large data collections (e.g., a groundwater database) and their components (e.g., the data sets corresponding to particular measuring stations). This approach to identifying objects is unsatisfactory for a variety of reasons. Most importantly, UDK objects may lose their identity when they are relocated in the primary tree due to some reorganization (such as the transfer of a department from one ministry to another). In this case, the objects that were relocated have to be recreated under a new ID at the new location. Newer versions of the UDK therefore resort to the OID concept described in Sect. 3.4.1. OIDs are created by the system. They are permanent and universally unique.

### 5.4.2 Object Classes and Inheritance

In order to structure the wide variety of UDK objects, and to facilitate both their capture and their administration, Günther et al. present a class concept [GLS96]. They distinguish seven classes of environmental data objects:

1. *Project data*: construction projects, environmental impact studies, etc.
2. *Empirical data*: measuring series, laboratory data, etc.
3. *Data about facilities*: factories, buildings, etc.
4. *Maps*
5. *Expertises and reports*
6. *Product data*
7. *Model data*: simulations etc.

For each of these seven classes there is a UDK class that contains the describing UDK objects. Each UDK class corresponds to a screen layout that is used for the capture and administration of the corresponding UDK objects. This pragmatic proposal was based on the user requirements that came up during the first months of UDK data capture. Obviously, this classification needs to be reviewed and possibly extended from time to time in order to reflect changing user requirements.

Another possible extension concerns the *vertical* structure of this classification. In particular, one could turn this flat class structure into an object-oriented class hierarchy that allows the inheritance of object attributes. The hierarchy could be structured as follows.

- The root of the hierarchy (level 0) consists of the generic class *UDK_Object* with four obligatory attributes: the unique object identifier (OID), the object name, the date when the object was last modified, and the agency (or the person) responsible for the object. Optional attributes, such as a textual description, may be included as well. Note that this generic class is not an abstract class, i.e., it may contain objects that are not included in any of its subclasses.
- Level 1 contains a relatively small number of classes that represent a consensus between all UDK participants. Currently, this level would correspond to the seven classes described above. Changes at this level are subject to negotiation between the UDK member countries.
- On subsequent levels of the hierarchy, participating countries or agencies are free to introduce additional subclasses depending on their particular requirements. This kind of flexibility is important not only for efficiency reasons. It is also crucial in order to secure acceptance for the UDK throughout its intended user community, especially in government agencies at the national and the local level.

Class attributes are inherited along this class hierarchy in an object-oriented manner. This includes the possibility to upgrade selected attributes from being optional to being required. It also means that attributes that are specific to a certain subclass, but not to its superclass(es), can be masked out when looking only at the superclass. For example, consider a particular topographic map $m$ and its UDK object $U_m$: $m$ is an element of the class *topographic_map*, which is a subclass of the class *map*. If one now looks at the UDK object $U_m$ through the screen layout corresponding to the class *map*, one only sees the attributes of *map*. The additional attributes that may have been introduced to describe *topographic* maps, as opposed to general maps, are not visible in this case.

This feature, which is typical for object-oriented environments, is a crucial element of standardization in the presence of application-specific extensions on the class hierarchy levels 2 and below. Any tool that is supposed to work at the national (or international) level across particular agencies or user communities can rely on the availability of the attributes defined at level 1. Maintenance and version management are other issues that need to rely on a stable class and attribute structure at the higher levels of the object hierarchy. It is therefore important to take organizational and technical precautions to make sure that users observe this principle in the presence of user-specific extensions and the resulting complexity in the class structure. The technical and organizational issues relating to these lower hierarchy levels are still subject to discussion.

### 5.4.3 Semantic Associations

Orthogonal to the class hierarchy described in the previous section, the UDK offers users the ability to connect *concrete UDK objects* with each other in a hypertext fashion. The resulting structures are directed graphs whose nodes correspond to UDK objects and whose edges represent semantic associations between them or between their respective environmental data objects. The semantics of those edges may vary. Note that those semantic nets are completely independent of the class hierarchy described in the previous section. While the nodes of the class hierarchy are *UDK object classes*, the nodes of the structures described in the following represent *concrete UDK objects*.

The most important such graph structure is the *primary tree* or *primary catalog*. Each UDK object corresponds to exactly one node of this tree structure, i.e., there is a 1:1 relationship between primary tree nodes and UDK objects. The links in the upper part of the tree serve to represent responsibilities of agencies and departments for particular sets of UDK objects. The agency that is in charge of a UDK object has to make sure that its information is correct and up to date. It is also responsible for the creation and deletion of UDK objects in the associated subtree(s). In the lower part of the tree, the links are used to represent *part-of* relationships between large data collections (e.g., a groundwater database) and their components (e.g., the data sets corresponding to particular measuring stations). An example is given in Fig. 5.11 depicting the UDK objects related to a groundwater database. Here the solid arrows make up the primary tree; their semantics varies between *is-responsible-for* (in the upper part of the tree) and *is-an-aggregation-of* (in the lower part).

Depending on particular user requirements, there may also be *secondary catalogs* to represent other semantic associations. Like the primary tree, a secondary catalog is a directed graph whose nodes each correspond to exactly one UDK object. Unlike the primary tree, however, the resulting structure no longer has to be a tree. Note also that a UDK object can be referenced by any number of secondary catalogs. There is a $1:n$ relationship between UDK objects and secondary catalog nodes: Each UDK object can be a node in any number of secondary catalogs, but each secondary catalog node refers to exactly one UDK object.

A typical application of secondary catalogs concerns the representation of additional aggregation relationships that are not represented in the primary tree. In Fig. 5.11 these kinds of associations are pictured as dotted arrows. These links are often useful for referring users first to relevant aggregated data sets before, upon request, giving them access to more detailed data. Another application of secondary catalogs is the construction of personal association structures. The *debate* association in Fig. 5.11 (dashed line) is an example of such a structure. For these free structures the system does not require users to restrict themselves to a tree structure. Like the freedom one has for linking pages in the World Wide Web, any directed graph structure is permitted, including graphs with cycles. The idea is to give UDK users a maximum amount of flexibility to connect and associate the various information items making up their working environment. With an attractive user interface, this option should be of great interest to a large group of users. What is important is that it has to be reasonably easy to create personal UDK objects and links. Furthermore, it is essential that those "personal" structures can be isolated from the public part of the UDK, so users can build confidential structures that are visible just for them or for their team.

In summary, it is important to note that the links connecting UDK objects may have a great variety of semantics. These different types of links need to be made explicit in the UDK by a labeling scheme. Users should have the option to choose the types of links they want to see at a given time. This would allow them to see a UDK object in a variety of contexts and to switch back and forth between those different representations. On the screen this could be supported, for example, by different colors and drawing modes for different types of links (Fig. 5.11).

### 5.4.4 Outlook

The UDK is a metainformation system and navigation tool that documents collections of environmental data from the government and other sources. Given the extreme success of the World Wide Web, we expect a significant amount of this kind of data to be available via WWW in the very near

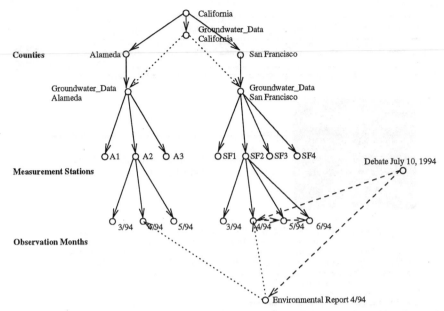

**Fig. 5.11.** Associations between UDK objects relating to groundwater data [GV98]

future. At this point there is no question that the Web is the most promising option to follow the spirit of the EU guideline and to make environmental information available to anybody who is interested. The UDK could play a major role in helping users to navigate in this overwhelming information pool, to identify which data is relevant for a given query, and to retrieve it fast and in a user-friendly manner.

Austria and several German states have released WWW implementations of the UDK (see, for example, http://udk.bmu.gv.at and http://www.uis-extern.um.bwl.de). Access is mainly keyword-based. The result of a search is a list of relevant UDK objects. More details on a particular object are available by checking it and sending the marked-up form back to the server. A CGI script then retrieves the corresponding additional attributes. Partly due to backlogs in data entry, however, most UDK objects are much less elaborate than the detailed example given in Fig. 5.10. Figures 5.12–5.15 show several pages from the Austrian UDK Web site.

HTML links between UDK objects or to environmental data objects are rarely used in those implementations. Instead one can request the ances-tor and descendants of a given UDK object in the primary catalog. This is done by means of another form-based mark-up mechanism, similar to the one described above. Further details of the current implementation and related issues have been described by Kramer et al. [KKN+96, KQ96, KNK+97].

A somewhat different approach for a WWW implementation of the UDK was suggested in [GLS96]. Here, each UDK object corresponds to exactly

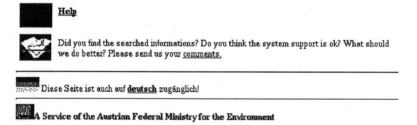

| File | Edit | View | Go | Bookmarks | Options | Directory | Window | Help |

Location: |http://udk.bmu.gv.at/cgi-bin/WWW-UDK/umlenkung.sh/welcome_a.sh

# WWW-UDK

## Welcome to the Austrian Environmental Data Catalogue!

The environmental data catalogue (UDK) by means of a meta-database substantial information about environmental data with regard to their subject, the geographical area concerned and their time-related description. These informations are compiled and administered by Austrian authorities and institutions covering also information about the latter. WWW-UDK is a tool to fulfill the requirements of the Environmental Information Act (UIG) and the Council Regulation Free Access to Environmental Information (90/313/EWR) of the Commission of the European Communities.

The UDK is aimed to be used by interested citizens and especially by employees in the environment-related administration units and in the field of environmental research.
The UDK is updated regularly in accordance with above mentioned institutions as well as federal and local authorities.
The further development of this innovative and strategic guidance tool will be realized in cooperation with the respective authorities in Germany, Switzerland, Liechtenstein and the European Environment Agency.

Please choose between the following information services:

**Search for adresses of authorities:**
- simple search for addresses
- search for tasks

**Search for available environmental data:**
- simple search
- complex search

**Help**

Did you find the searched informations? Do you think the system support is ok? What should we do better? Please send us your comments.

Diese Seite ist auch auf **deutsch** zugänglich!

A Service of the **Austrian Federal Ministry for the Environment**

**Fig. 5.12.** Entry page of the Austrian UDK

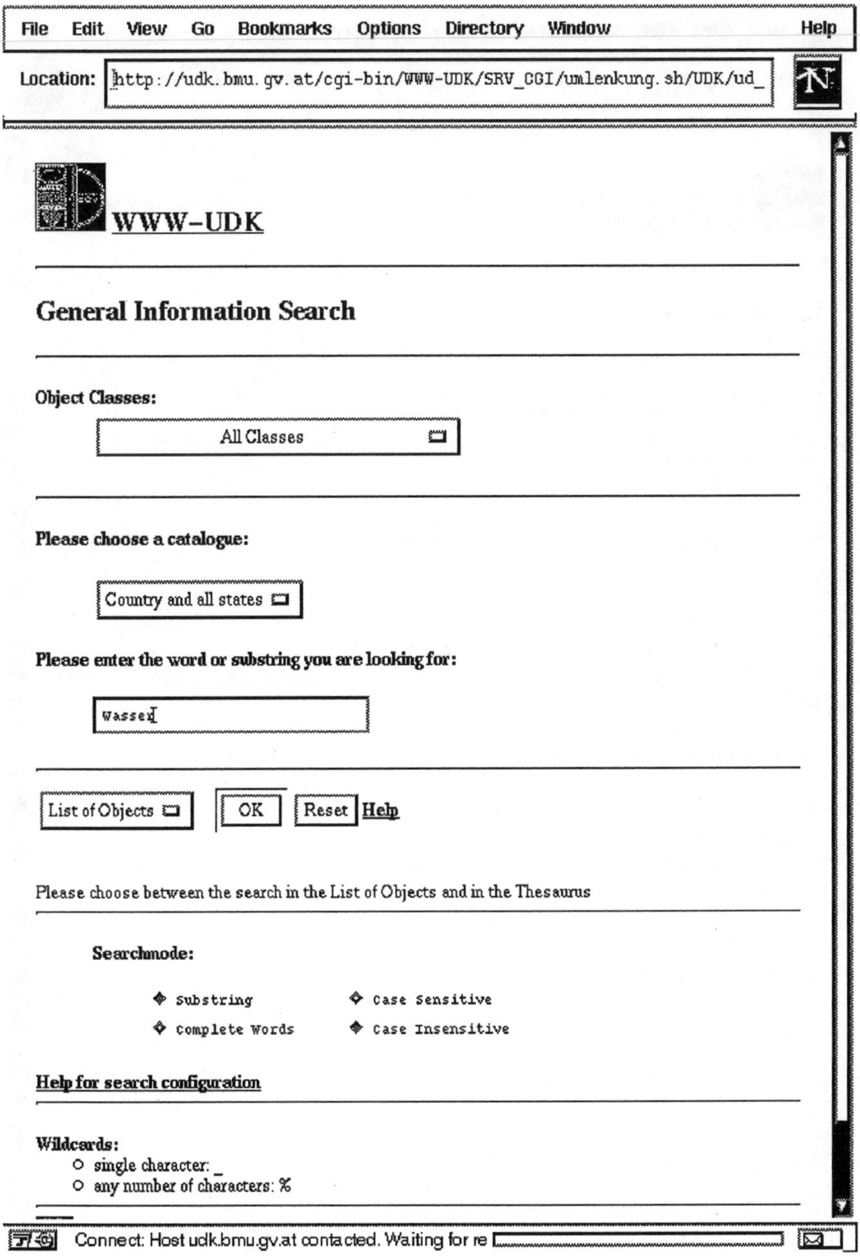

| File | Edit | View | Go | Bookmarks | Options | Directory | Window | | Help |

Location: | http://udk.bmu.gv.at/cgi-bin/WWW-UDK/SRV_CGI/umlenkung.sh/UDK/ud_ | N

**WWW–UDK**

## General Information Search

**Object Classes:**

| All Classes ▭ |

**Please choose a catalogue:**

| Country and all states ▭ |

**Please enter the word or substring you are looking for:**

| wasse| |

| List of Objects ▭ |    | OK |  | Reset | **Help** |

Please choose between the search in the List of Objects and in the Thesaurus

**Searchmode:**

♦ Substring          ♦ Case Sensitive

♦ Complete Words     ♦ Case Insensitive

**Help for search configuration**

**Wildcards:**
○ single character: _
○ any number of characters: %

⌨🔊 Connect: Host udk.bmu.gv.at contacted. Waiting for re [ ▭ ] ✉

**Fig. 5.13.** Searching the UDK records

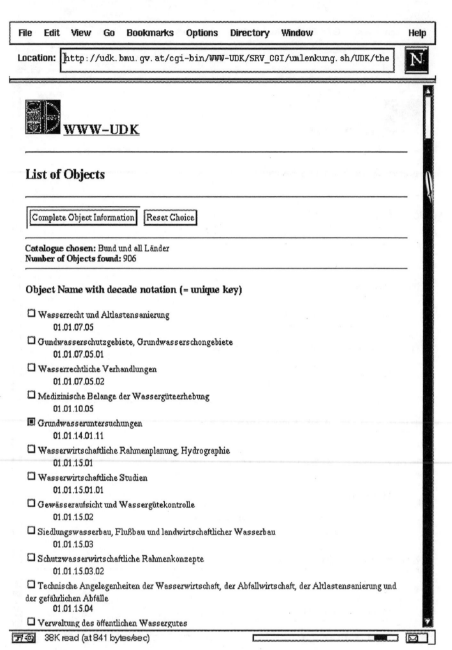

**Fig. 5.14.** UDK search results for the term *Wasser (water)*

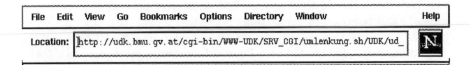

| File | Edit | View | Go | Bookmarks | Options | Directory | Window | | Help |

Location: [http://udk.bmu.gv.at/cgi-bin/WWW-UDK/SRV_CGI/umlenkung.sh/UDK/ud_    N.

 **WWW–UDK**

## Complete Object Information

### 1. EDC–Object

| General Information | |
|---|---|
| **Object Name:** | Grundwasseruntersuchungen |
| **EDC–Class:** | Plan/Projects/Programs |
| **Decadic Notation:** | 01.01.14.01.11 |
| **Free Search Terms:** | |
| **Thesaurus Search Terms:** | |
| **Address Lable:** | **AUTOELRBGLDXIII/20** |

**List of Predecessors and Successors:**

**Predecessors:**

⊞Technische Angelegenheiten des Straßenbaues, insbesondere Projektierung, Bau und Erhaltung von Landesstraßen und Bundesstraßen

| Predecessors/Successors | | Reset Choice |

A Service of the **Austrian Federal Ministry for the Environment, Youth, and Family**

**Fig. 5.15.** UDK catalog entry

one Web page. HTML links are used to implement primary and secondary catalogs and to establish connections to environmental data objects. A corresponding realization is currently under consideration for VKS, a metainformation system for the German Federal Environmental Agency [SMJ97].

Besides stand-alone implementations of the UDK, the UDK data sets are also used in other contexts to improve accessibility to the underlying data sources. For example, UDK data sets are included in the index facility (the *locator*) of the German Environmental Information Network (GEIN, cf. Sect. 4.6.5). If available, the search result includes a WWW link to the actual data.

## 5.5 Summary

The goal of this chapter has been to show how metadata is becoming increasingly popular in environmental information systems, in order to improve both the availability and the quality of the information delivered. The growing popularity of Internet-based data servers has accelerated this trend even further. After a general discussion of the term *metadata* and of the possibility to integrate metadata into traditional information system architectures, we have discussed several case studies in detail. Particular emphasis has been put on the U.S. efforts to build a National Spatial Data Infrastructure, and on several European projects to integrate environmental information processing at the national and international level.

Despite the remaining heterogeneities and inefficiencies, the outlook seems positive. The ubiquitous trend towards open systems as well as the rise of the World Wide Web are two recent developments that will greatly improve the way we manage environmental information. Users will have faster and more comfortable access to ever greater amounts of information, and metadata will be an essential component of the underlying software architectures.

Finally, we envisage an increasing number of applications where metadata is used to administer not only simple data sets but also complex software tools, such as domain-specific aggregation methods or environmental simulation models. In those applications, the metadata will be used for two purposes: (i) to find the appropriate software tool for a given problem, and (ii) to to apply the tool to a given data set over the Internet without having to port the software to a local machine. Our own MMM project [GMS+97] is one example of a software architecture that supports this paradigm.

# 6. Conclusions

Environmental information systems are one of the most challenging applications areas for computer scientists. Due to the volume and complexity of
environmental data, users require state-of-the-art techniques from a variety
of disciplines. They are often willing to try out new approaches, and as a
result, cooperations between users, researchers, and vendors are unusually
intense.

Statistical classification, possibly enhanced by modern approaches to uncertainty management, is used to perform the initial filtering. Terabytes of
seemingly unstructured raw data are transformed into formatted data sets
that represent a semantic aggregation of the given inputs. Latest database
technology is then needed to store and query those data sets. Recent advances
in spatial databases and geographic information systems provide the means
for doing so efficiently. To prepare the data for decision support purposes, one
requires sophisticated analysis tools, such as simulation models, information
retrieval techniques, and visualization packages. Systems for the management
of associated metadata provide crucial support for searching and processing
environmental information.

Throughout this information flow, data is aggregated and compressed to
serve the needs of decision makers. In this book we have tried to cover the
base techniques underlying this complex aggregation process. Our treatment
is structured into four major parts: data capture, data storage, data analysis,
and metadata management. Because of the multitude of disciplines involved
we were not able to discuss every technique in detail. References to more
specialized literature are provided for the interested reader.

In addition to our discussion of the *concepts* underlying environmental
information management, we presented several *case studies* of applied research projects and of environmental information systems in practice. As
most of these systems are under continuous development, one should not be
surprised to see minor differences between the descriptions given here and
the functionalities at the time this book appears in print. For more detailed
and up-to-date documentation, readers are therefore advised to contact the
responsible agencies.

Looking at the field of environmental informatics in a more general sense,
the breadth and depth of related activities is still increasing. Besides a large

number of recent publications in journals and conference proceedings, there have been comprehensive anthologies in English [AP95] and German [PH95]. In addition, the various communities involved in such projects are starting to meet and to conduct joint activities. In the past, the field has been somewhat fragmented along both national and subject borders.

Germany and Austria, for example, have a long history of related research and development activities. Since 1987, the German *Gesellschaft für Informatik* has organized an annual symposium on environmental informatics; see [JKPR93, HJPS94, KP95, LL96, GJR+97] for recent proceedings. While the first conferences were mostly local, more recent events were advertised internationally and involved an increasing number of colleagues from the United States, France, and other countries. In 1997, the symposium was held for the first time in a non-German speaking country (France).

North American efforts do not suffer from the same language barriers. On the other hand, there have traditionally been few activities that transcend the traditional subject boundaries between the various disciplines that conduct environmentally relevant research. For example, researchers in the geosciences and environmental sciences have long been working on adequate computer support for their core activities. Commercial geographic information systems are among the most important outcomes of these efforts. Computer scientists have been involved only marginally in the development of such systems. More generally speaking, cooperations between computer scientists and the environmental science community have been rare until recently.

Since the early 1990s, however, the increasing number of contacts is bearing fruit. There have been several joint projects of high visibility, including the National Center for Geographic Information and Analysis [Nat98] and the Sequoia 2000 project [Sto93, Fre94]. U.C. Berkeley's Digital Environmental Library (ELIB) project [Uni98c, Wil96], and U.C. Santa Barbara's Alexandria project [Smi96] are pursuing similar goals; both projects are funded through the NSF/ARPA/NASA Digital Library Initiative. Since 1993, there has also been the bi-annual ISESS symposium series on environmental software systems [DSR95, DSS97].

These and related activities have resulted in numerous joint publications and conferences, as well as system solutions and commercial products. Besides the proceedings of the conferences listed above, good starting points into the related literature include the collections edited by Goodchild et al. [GPS93, GSP96], Michener et al. [MBS94], Avouris and Page [AP95], and Günther [Gün97].

We certainly hope that this book contributes to making the area of environmental information systems known to a broader audience, establishing the subject in the academic curriculum, and increasing the number of researchers and practitioners working on related problems. It is after all an exceptionally attractive opportunity for computer scientists and engineers to contribute directly to one of the great challenges of our time: to maintain and protect our natural environment in an era of unprecedented population and industrial growth. While technical progress has been a key factor in the developments that led to the current environmental crisis, it is now essential to turn to technology once again to help us solve those problems. Inventing and developing computer technology for environmental applications is one important cornerstone of this endeavor.

# List of URLs[1]

AlfaWeb Hazardous Substance Information System
      www.iai.fzk.de/~weidemann/lfu/lfu.htm
A M Productions
      www.amproductions.com/contentg.html
ASK – Global Change Directory of Information Services
      ask.gcdis.usgcrp.gov:8080
Bay Area Geodata
      badger.parl.com
Brown is Green Resource Conservation Program
      www.brown.edu/Departments/Brown_Is_Green
Catalogue of Data Sources
      www.mu.niedersachsen.de/cds/webcds
Center for Earth Observation
      www.ceo.org
Central European Environmental Data Request Facility
      www.cedar.univie.ac.at
CHEMTOX
      www.krinfo.com/dialog/databases/html2.0/bl0337.html
CIESIN Demographic Data Viewer
      plue.sedac.ciesin.org/plue/ddviewer
CIESIN Remote Sensing Thematic Guide
      www.ciesin.org/TG/RS/RS-home.html
Common Gateway Interface
      hoohoo.ncsa.uiuc.edu/cgi/overview.html
Cygnus Group – Integration of Environmental and Business Concepts
      www.cygnus-group.com
Daily Planet
      www.atmos.uiuc.edu
DAIN – Internet Resources on Environmental Protection
      dino.wiz.uni-kassel.de/dain.html

---

[1] All URLs in this list follow the Hypertext Transfer Protocol, i.e., they have
to be prefixed by http://. An online version of this list can be found at
http://www.wiwi.hu-berlin.de/~guenther/eis_book.html. All URLs have been
verified between October 1997 and January 1998.

Data-Star
    www.krinfo.com/dialog/publications/data-star-mini-catalogue.html
Dialog
    www.dialog.com
Earth Observation Sciences
    www.eos.co.uk
Earth Observing System Data and Information System
    spsosun.gsfc.nasa.gov/EOSDIS_main.html
Earth Pages
    starsky.hitc.com/earth/earth.html
EcoWeb
    ecosys.drdr.virginia.edu/EcoWeb.html
Enviroline
    www.krinfo.com/dialog/databases/html2.0/bl031.html
Envirolink
    www.envirolink.org
Environmental News Network
    www.enn.com
Environmental Systems Research Institute
    www.esri.com
European Commission Host Organisation
    www.echo.lu
European Environment Agency
    www.eea.dk
European Science Foundation – GISDATA
    www.shef.ac.uk/uni/academic/D-H/gis/gisdata.html
EWSE – An Earth Observation Information Exchange
    ewse.ceo.org
G7 Environment and Natual Resources Management
    enrm.ceo.org
German Environmental Information Network
    www.faw.uni-ulm.de/deutsch/Literatur/Radermacher/ministry.html
German Federal Environmental Agency
    www.umweltbundesamt.de
GIS Courses Online
    www.frw.ruu.nl/eurogi
Global Change Master Directory
    gcmd.gsfc.nasa.gov
Global Network for Environmental Technology
    gnet.together.org
Grasslinks
    www.regis.berkeley.edu/grasslinks
IEEE Metadata Homepage
    www.llnl.gov/liv_comp/metadata

Informix
　　　www.informix.com
Institute for Scientific Organization
　　　www.isinet.com
IRIS Visual Data Mapping System
　　　allanon.gmd.de/and/and.html
Java
　　　java.sun.com
Java-based ESRI Shapefile Viewer
　　　www.gis.umn.edu/fornet/java/shpclient
J&W Scientific
　　　www.jandw.com
Landsat Pathfinder
　　　amazon.sr.unh.edu/pathfinder1/index.html
Landsat TM Data for Minnesota
　　　www.gis.umn.edu/fornet/ids/imageview
MEGRIN Surveying and Mapping Data
　　　www.ign.fr/megrin/megrin.html
National Center for Geographic Information and Analysis
　　　www.ncgia.ucsb.edu
National Environmental Data Index Catalog
　　　www.nedi.gov/NEDI-Catalog
National Environmental Satellite Data and Inf. System
　　　ns.noaa.gov/NESDIS/NESDIS_Home.html
NCGIA Core Curriculum
　　　www.ncgia.ucsb.edu/education/ed.html
New South Wales Environment Protection Agency
　　　www.epa.nsw.gov.au
ObjectStore
　　　www.objectstore.com
Open GIS Consortium
　　　www.opengis.org
PDLCOM Online Database
　　　info.cas.org/ONLINE/DBSS/pdlcomss.html
Register of Ecological Models
　　　dino.wiz.uni-kassel.de/ecobas.html
SAP AG
　　　www.sap.com
Scientific Computers GmbH
　　　www.scientific.de
Siemens Nixdorf Informationssysteme AG
　　　www.sni.de
Smallworld
　　　www.smallworld-us.com

SNIG Cover Page
        snig.cnig.pt
Spatial Data Transfer Standard
        mcmcweb.er.usgs.gov/sdts
SPOT Satellite System
        www.spot.com
SPOT Image
        www.spotimage.fr/anglaise/system/s_syst.htm
STN
        www.cas.org/stn.html
TELSAT Guide for Satellite Imagery
        www.belspo.be/telsat
U.C. Berkeley Digital Library Project
        elib.cs.berkeley.edu
UDK Austria
        udk.bmu.gv.at
UFORDAT
        info.cas.org/ONLINE/DBSS/ufordatss.html
UIS Baden-Württemberg
        www.uis-extern.um.bwl.de
UNESCO Environmental Programs
        www.unesco.org/general/eng/programmes/science/index.html
UNIGIS
        www.unigis.org
U.S. Bureau of the Census
        www.census.gov
U.S. Census Tiger Files
        www.census.gov/geo/www/tiger
U.S. Environmental Protection Agency
        www.epa.gov
U.S. Federal Geographic Data Committee
        fgdc.er.usgs.gov
U.S. Geological Survey
        info.er.usgs.gov
U.S. Government Information Locator Service
        www.epa.gov/gils
U.S. National Geospatial Data Clearinghouse
        nsdi.usgs.gov/nsdi
WASY GmbH
        www.wasy.de
WWW Virtual Library Environment
        earthsystems.org/Environment.shtml

# References

[AB96] L. Anselin and S. Bao. Exploratory spatial data analysis linking SpaceStat and ArcView. Technical report, West Virginia University, Morgantown, WVa., 1996.

[ABD+89] M. Atkinson, F. Bancilhon, D. DeWitt, K. Dittrich, D. Maier, and S. Zdonik. The object-oriented database system manifesto. In *Proc. First Int. Conf. on Deductive and Object-Oriented Databases*, 1989. Reprinted in [Sto94].

[Abe97] D. J. Abel. Spatial Internet marketplaces: A grand challenge. In M. Scholl and A. Voisard, editors, *Advances in Spatial Databases*. LNCS 1262. Springer-Verlag, Berlin/Heidelberg/New York, 1997.

[Ada84] J. Barclay Adams. Probabilistic reasoning and certainty factors. In B. Buchanan and E. Shortliffe, editors, *Rule Based Expert Systems*, pages 263–271. Addison-Wesley, Reading, Mass., 1984.

[AG94] H.-K. Arndt and O. Günther. Umwelthaftung – Anforderungen an das Umwelt-Controlling und an betriebliche Umweltinformationssysteme. In L. M. Hilty, A. Jaeschke, B. Page, and A. Schwabl, editors, *Informatik für den Umweltschutz*. Metropolis, Marburg, Germany, 1994. Vol. 2.

[AG96] H.-K. Arndt and O. Günther. Betriebliche Umweltinformationssysteme. *UmweltWirtschaftsForum*, 4(1), 1996.

[AG97] N. R. Adam and A. Gangopadhyay, editors. *Database Issues in Geographic Information Systems*. Kluwer Academic Publishers, Norwell, Mass., 1997.

[AGHR97] H.-K. Arndt, O. Günther, L. M. Hilty, and C. Rautenstrauch, editors. *Metainformationen und Datenintegration in betrieblichen Umweltinformationssystemen (BUIS)*. Metropolis, Marburg, Germany, 1997.

[AM90] D. J. Abel and D. M. Mark. A comparative analysis of some two-dimensional orderings. *Int. J. Geographical Information Systems*, 4(1):21–31, 1990.

[AO93] D. Abel and B. C. Ooi, editors. *Advances in Spatial Databases*. LNCS 692. Springer-Verlag, Berlin/Heidelberg/New York, 1993.

[AP95] N. M. Avouris and B. Page, editors. *Environmental Informatics – Methodology and Applications of Environmental Information Processing*. Kluwer Academic Publishers, Norwell, Mass., 1995.

[Aro89] S. Aronoff. *Geographic Information Systems: A Management Perspective*. WDL Publications, Ottawa, 1989.

[AS83] D. J. Abel and J. L. Smith. A data structure and algorithm based on a linear key for a rectangle retrieval problem. *Computer Vision*, 24:1–13, 1983.

[AS94] W. G. Aref and H. Samet. The spatial filter revisited. In *Proc. 6th Int. Symp. on Spatial Data Handling*, pages 190–208, 1994.

[ATK97] D. J. Abel, K. Taylor, and D. Kuo. Integrating modelling systems for environmental management information systems. *SIGMOD Record*, 26(1):5–10, 1997.

[AYA+92] D. Abel, K. Yap, R. Ackland, M. Cameron, D. Smith, and G. Walker. Environmental decision support system project: An exploration of alternative architectures for geographic information systems. *Int. J. Geographical Information Systems*, 6(3):193–204, 1992.

[Bar95] J. G. Bartzis. Environmental monitoring and simulation. In N. M. Avouris and B. Page, editors, *Environmental Informatics – Methodology and Applications of Environmental Information Processing*, chapter 15, pages 237–256. Kluwer Academic Publishers, Norwell, Mass., 1995.

[Bat94] W.-D. Batschi. Environmental thesaurus and classification of the Umweltbundesamt (German Federal Environmental Agency). In P. Stancikova and I. Dahlberg, editors, *Environmental Knowledge Organization and Information Management*. INDEKS, Frankfurt/Main, 1994.

[Bay96] R. Bayer. The Universal B-tree for multidimensional indexing. Technical Report I9639, Technische Universität, Munich, Germany, 1996. http://www.leo.org/pub/comp/doc/techreports/tum/informatik/report/1996/TUM-I9639.ps.gz.

[BC94] G. Bonham-Carter. *Geographic Information Systems: Modeling with GIS*. Pergamon, New York, 1994.

[BDC+95] C. Bardinet, J. E. Dubois, J. P. Caliste, J. J. Royer, and J. C. Oppeneau. Data processing for the environment analysis: A multiscale approach. In *Space and Time in Environmental Information Systems*. Metropolis, Marburg, Germany, 1995.

[BDK92] F. Bancilhon, C. Delobel, and P. Kannelakis, editors. *Building an Object-Oriented Database System*. Morgan Kaufmann, San Mateo, Calif., 1992.

[Bec92] L. Becker. *A New Algorithm and a Cost Model for Join Processing with the Grid File*. PhD thesis, Universität-Gesamthochschule Siegen, Germany, 1992.

[Bel95] Belgian Federal Government. *The TELSAT Guide for Satellite Imagery*. Federal Office for Scientific, Technical and Cultural Affairs, 1995. URL http://www.belspo.be/telsat.

[Ben75] J. L. Bentley. Multidimensional binary search trees used for associative searching. *Comm. ACM*, 18(9):509–517, 1975.

[Ben79] J. L. Bentley. Multidimensional binary search in database applications. *IEEE Trans. Software Eng.*, 4(5):333–340, 1979.

[BF79] J. L. Bentley and J. H. Friedman. Data structures for range searching. *ACM Comp. Surv.*, 11(4):397–409, 1979.

[BF95] A. Belussi and C. Faloutsos. Estimating the selectivity of spatial queries using the "correlation" fractal dimension. In *Proc. 21st Int. Conf. on Very Large Data Bases*, pages 299–310, 1995.

[BF97] R. Bill and D. Fritsch. *Grundlagen der Geoinformationssysteme I*. Wichmann, Karlsruhe, Germany, 4th edition, 1997.

[BG92] L. Becker and R. H. Güting. Rule-based optimization and query processing in an extensible geometric database system. *ACM Trans. Database Systems*, 17(2):247–303, 1992.

[BGSW90] A. Buchmann, O. Günther, T. R. Smith, and Y.-F. Wang. *Design and Implementation of Large Spatial Databases*. LNCS 409. Springer-Verlag, Berlin/Heidelberg/New York, 1990.

[Bil95] R. Bill. Spatial data processing in environmental information systems. In N. M. Avouris and B. Page, editors, *Environmental Informatics – Methodology and Applications of Environmental Information Processing*, chapter 5, pages 53–74. Kluwer Academic Publishers, Norwell, Mass., 1995.

[Bil96] R. Bill. *Grundlagen der Geoinformationssysteme II*. Wichmann, Karlsruhe, Germany, 2nd edition, 1996.

[BK94]  T. Brinkhoff and H.-P. Kriegel. The impact of global clustering on spatial database systems. In *Proc. 20th Int. Conf. on Very Large Data Bases*, pages 168–179, 1994.

[BKS93]  T. Brinkhoff, H.-P. Kriegel, and B. Seeger. Efficient processing of spatial joins using R-trees. In *Proc. ACM SIGMOD Int. Conf. on Management of Data*, pages 237–246, 1993.

[BKSS90]  N. Beckmann, H.-P. Kriegel, R. Schneider, and B. Seeger. The R*-tree: An efficient and robust access method for points and rectangles. In *Proc. ACM SIGMOD Int. Conf. on Management of Data*, pages 322–331, 1990.

[BKSS94]  T. Brinkhoff, H.-P. Kriegel, R. Schneider, and B. Seeger. Multi-step processing of spatial joins. In *Proc. ACM SIGMOD Int. Conf. on Management of Data*, pages 197–208, 1994.

[BM72]  R. Bayer and E. M. McCreight. Organization and maintenance of large ordered indices. *Acta Informatica*, 1(3):173–189, 1972.

[BMW93]  A. Braunschweig and R. Müller-Wenk. *Ökobilanzen für Unternehmen.* Verlag Paul Haupt, Bern/Stuttgart/Wien, 1993.

[BN97]  P. Batty and R. G. Newell. GIS databases are different, 1997. URL http://www.smallworld-us.com.

[Bos94]  H. Bossel. *Modeling and Simulation.* A. K. Peters, Wellesley, Mass., 1994.

[Bri92]  British Standard Institution. Specification for environmental management systems (BS 7750). London, 1992.

[BT97]  H. K. Bhargava and C. G. Tettelbach. A Web-based decision support system for waste disposal and recycling. *Computers, Environment, and Urban Systems*, 21, 1997.

[Bur84]  W. A. Burkhard. Index maintenance for non-uniform record distributions. In *Proc. ACM SIGACT/SIGMOD Symp. on Principles of Database Systems*, pages 173–180, 1984.

[Bur86]  P. A. Burrough. *Principles of Geographical Information Systems for Land Resources Assessment.* Clarendon Press, Oxford, 1986.

[BV95]  J. Benz and K. Voigt. Umwelt-Metadatenbanken im Internet. In B. Page and L. M. Hilty, editors, *Umweltinformatik – Informatikmethoden für Umweltschutz und Umweltforschung.* Oldenbourg, Munich/Vienna, 2nd edition, 1995.

[BV96]  J. Benz and K. Voigt. Aufbau eines Systems zur strukturierten Suche nach Informationsquellen für den Umweltschutz im Internet. In H. Lessing and U. Lipeck, editors, *Informatik im Umweltschutz.* Metropolis, Marburg, Germany, 1996.

[CAP94]  CAP debis. *debis-UIS Umweltinformationssystem.* Fellbach, Germany, 1994.

[CC90]  W. J. Campbell and R. F. Cromp. Evolution of an intelligent information fusion system. *Photogrammetric Engineering and Remote Sensing*, 56(6):867–870, 1990.

[CD97]  S. Chaudhuri and U. Dayal. An overview of data warehousing and OLAP technology. *SIGMOD Record*, 26(1):65–74, 1997.

[CDHH94]  E. Czorny, W. Dresselhaus, D. Haas, and B.-P. Hamels. EXCEPT: Symbiose aus Forschung, Anwendungsentwicklung und Anwendern. In L. M. Hilty, A. Jaeschke, B. Page, and A. Schwabl, editors, *Informatik für den Umweltschutz.* Metropolis, Marburg, Germany, 1994. Vol. 1.

[CDK+95]  L. T. Chen, R. Drach, M. Keating, S. Louis, D. Rotem, and A. Shoshani. Access to multidimensional datasets on tertiary storage systems. *Information Systems*, 20(2):155–183, 1995.

[CKM93] M. Clarke, R. Kruse, and S. Moral, editors. *Symbolic and Quantitative Approaches to Reasoning and Uncertainty.* LNCS 747. Springer-Verlag, Berlin/Heidelberg/New York, 1993.

[Coc77] W. G. Cochran. *Sampling Techniques.* John Wiley & Sons, New York, 1977.

[Com79] D. Comer. The ubiquitous B-tree. *ACM Comp. Surv.*, 11(2):121–138, 1979.

[Cou90] Council of the European Communities. Council Directive (90/313/EEC) of 7 June 1990 on the freedom of access to information on the environment. *Official Journal of the European Communities*, L158:56–58, 1990.

[Cou93] Council of the European Communities. Council Regulation (93/ 1836/EEC) of 29 June 1993 allowing participation by companies in the industrial sector in a Community eco-management and audit scheme. *Official Journal of the European Communities*, L168:1–18, 1993.

[CSMC97] D. Cook, J. Symanzik, J. J. Majure, and N. Cressie. Dynamic graphics in a GIS: More examples using linked software. *Computers and Geosciences*, 23(4), 1997.

[Dav91] F. W. Davis et al. Environmental analysis using integrated GIS and remotely sensed data. *Photogrammetric Engineering and Remote Sensing*, 57(6):689–697, 1991.

[Dem67] A. P. Dempster. Upper and lower probabilities induced by a multivalued mapping. *Annals of Mathematical Statistics*, 38, 1967.

[Dem68] A. P. Dempster. A generalization of Bayesian inference. *Journal of the Royal Statistical Society*, 30:205–247, 1968.

[Den91] P. J. Densham. Spatial decision support systems. In D. J. Maguire, M. F. Goodchild, and D. W. Rhind, editors, *Geographical Information Systems: Principles and Applications*, volume 1, chapter 26, pages 403–412. Longman, Harlow, UK, 1991.

[Den93] R. Denzer. Concepts for the visual presentation of environmental data. In A. Jaeschke, T. Kämpke, B. Page, and F. J. Radermacher, editors, *Informatik im Umweltschutz*. Informatik aktuell. Springer-Verlag, Berlin/Heidelberg/New York, 1993.

[Den94] D. K. Denton. *Enviro-Management – How Smart Companies Turn Environmental Costs Into Profits.* Prentice-Hall, Englewood Cliffs, N.J., 1994.

[Des90] J. Desachy. ICARE: An expert system for automatic mapping from satellite imagery. In L. F. Pau, editor, *Mapping and Spatial Modeling for Navigation.* Springer-Verlag, Berlin/Heidelberg/New York, 1990.

[Deu90] O. Deux et al. The story of $O_2$. *IEEE Trans. Knowledge and Data Eng.*, 2(1):91–108, 1990.

[Deu91] O. Deux et al. The $O_2$ system. *Comm. ACM*, 34(10):34–48, October 1991.

[Dit86] K. R. Dittrich. Object-oriented database systems: The notion and the issues. In *Proc. 1986 Int. Workshop on Object-Oriented Database Systems*, Washington, DC, 1986. IEEE Computer Society Press.

[Dit90] K. R. Dittrich. Object-oriented database systems: The next miles of the marathon. *Information Systems*, 15(1):161–167, 1990.

[DMH95] R. Denzer, H. F. Mayer, and W. Haas. Visualization of environmental data. In N. M. Avouris and B. Page, editors, *Environmental Informatics – Methodology and Applications of Environmental Information Processing*, chapter 6, pages 75–92. Kluwer Academic Publishers, 1995.

[DPSB97] K. Doan, C. Plaisant, B. Shneiderman, and T. Bruns. Query previews for networked information systems: A case study with NASA environmental data. *SIGMOD Record*, 26(1):75–81, 1997.

[DSR95] R. Denzer, G. Schimak, and D. Russell, editors. *Environmental Software Systems*. Chapman and Hall, London, 1995.

[DSS97] R. Denzer, D. A. Swayne, and G. Schimak, editors. *Environmental Software Systems*. Chapman and Hall, London, 1997.

[DY97] P. Drew and J. Ying. Metadata management for geographic information discovery and exchange. In W. Klas and A. Sheth, editors, *Managing Multimedia Data: Using Metadata to Integrate and Apply Digital Data*. McGraw-Hill, New York, 1997.

[Edd93] J. A. Eddy. Environmental research: What we must do. In M. F. Goodchild, B. O. Parks, and L. T. Steyaert, editors, *Environmental Modeling With GIS*, chapter 1, pages 3–7. Oxford University Press, New York/Oxford, 1993.

[Ede85] H. Edelsbrunner. *Algorithms in Combinatorial Geometry*. Springer-Verlag, Berlin/Heidelberg/New York, 1985.

[EF88] M. Egenhofer and A. Frank. Towards a spatial query language: User interface considerations. In *Proc. 14th Int. Conf. on Very Large Data Bases*, pages 124–133, 1988.

[EFJ90] M. Egenhofer, A. U. Frank, and J. P. Jackson. A topological data model for spatial databases. In A. Buchmann, O. Günther, T. R. Smith, and Y.-F. Wang, editors, *Design and Implementation of Large Spatial Databases*. LNCS 409. Springer-Verlag, Berlin/Heidelberg/New York, 1990.

[Ege90] M. Egenhofer. A formal definition of binary topological relationships. In *Proc. 3rd Int. Conf. on Foundations of Data Organization and Algorithms*. LNCS 367, pages 322–338. Springer-Verlag, Berlin/Heidelberg/New York, 1990.

[Ege91] M. Egenhofer. Extending SQL for cartographic display. *Cartography and Geographic Information Systems*, 18:230–245, 1991.

[Ege92] M. Egenhofer. Why not SQL! *Int. J. Geographical Information Systems*, 6(2):71–85, 1992.

[Ege94] M. Egenhofer. Spatial SQL: A query and presentation language. *IEEE Trans. Knowledge and Data Eng.*, 6(1):86–95, 1994.

[EGSS91] M. Ehlers, D. Greenlee, T. Smith, and J. Star. Integration of remote sensing and GIS: Data and data access. *Photogrammetric Engineering and Remote Sensing*, 57(6):669–675, 1991.

[EH95] M. J. Egenhofer and J. R. Herring, editors. *Advances in Spatial Databases*. LNCS 951. Springer-Verlag, Berlin/Heidelberg/New York, 1995.

[ELS95] G. Evangelidis, D. Lomet, and B. Salzberg. The hB$^{\Pi}$-tree: A modified hB-tree supporting concurrency, recovery and node consolidation. In *Proc. 21st Int. Conf. on Very Large Data Bases*, pages 551–561, 1995.

[ESR91] ESRI Inc., editor. *Proc. 11th ESRI User Conference*. ESRI Inc., Redlands, Calif., 1991.

[ESR95a] ESRI Inc. Metadata management in GIS. Technical report, ESRI Inc., Redlands, Calif., 1995. URL http://www.esri.com/resources/papers/papers.html.

[ESR95b] ESRI Inc., editor. *Proc. 15th ESRI User Conference*. ESRI Inc., Redlands, Calif., 1995.

[ESR95c] ESRI Inc. SDTS – Supporting the Spatial Data Transfer Standard in ARC/INFO. Technical report, ESRI Inc., Redlands, Calif., 1995. URL http://www.esri.com/resources/papers/papers.html.

[ESR97a] ESRI Inc. ESRI proposes solutions to OpenGIS. *ARC News*, 18(4):21, 1997.

[ESR97b] ESRI Inc. GIS for everyone – now on Web! *ARC News*, 18(4):1–3, 1997.

[ESR97c] ESRI Inc. IBM and Informix select ESRI to spatially enable companies' respective technologies. *ARC News*, 18(4):9, 1997.

[ESR98] ESRI Inc., 1998. URL http://www.esri.com.

[Eur93] European Environment Agency. Catalogue of Datasources for the Environment. Copenhagen, 1993. Version 930831.

[Eur96] European Environment Agency. Catalogue of Data Sources and Thesaurus – Implementation Reference Document. Copenhagen, 1996. Draft Version 0.3. URL http://www.mu.niedersachsen.de/cds.

[Eur97a] European Committee for Standardization. Environmental Management – Life Cycle Assessment – Principles and Framework (ISI 14040:1997). Brussels, 1997.

[Eur97b] European Committee for Standardization. Environmental Management – Life Cycle Assessment – Goal and Scope Definition and Life Cycle Inventory Analysis (ISI 14041:1997). Brussels, 1997.

[Eur97c] European Committee for Standardization. Environmental Management – Life Cycle Assessment – Life Cycle Impact Assessment (ISI 14042:1997). Brussels, 1997.

[Eur97d] European Committee for Standardization. Environmental Management – Life Cycle Assessment – Life Cycle Interpretation (ISI 14043:1997). Brussels, 1997.

[Eur97e] European Topic Center on Catalogue of Data Sources. *ETC/CDS Newsletter*, 4, 1997.

[Eur98] European Topic Center on Catalogue of Data Sources. *CDS Home Page.* http://www.mu.niedersachsen.de, 1998.

[FAG83] H. Fuchs, G. D. Abram, and E. D. Grant. Near real-time shaded display of rigid objects. *Computer Graphics*, 17(3):65–72, 1983.

[Fal85] C. Faloutsos. Multiattribute hashing using Gray-codes. In *Proc. ACM SIGMOD Int. Conf. on Management of Data*, pages 227–238, 1985.

[Fal88] C. Faloutsos. Gray-codes for partial match and range queries. *IEEE Trans. Software Eng.*, 14:1381–1393, 1988.

[FB74] R. A. Finkel and J. L. Bentley. Quad trees: A data structure for retrieval of composite keys. *Acta Informatica*, 4(1):1–9, 1974.

[FB90] A. Frank and R. Barrera. The fieldtree: A data structure for geographic information systems. In A. Buchmann, O. Günther, T. R. Smith, and Y.-F. Wang, editors, *Design and Implementation of Large Spatial Databases*. LNCS 409, pages 29–44. Springer-Verlag, Berlin/Heidelberg/New York, 1990.

[FCL92] F. G. Fegeas, J. L. Cascio, and R. A. Lazar. An overview of FIPS 173, the Spatial Data Transfer Standard. *Cartography and Geographic Information Systems*, 19(5), 1992.

[FD97] J. Frew and J. Dozier. Data management for earth system science. *SIGMOD Record*, 26(1):27–31, 1997.

[Fed93] K. Fedra. GIS and environmental modeling. In M. F. Goodchild, B. O. Parks, and L. T. Steyaert, editors, *Environmental Modeling with GIS*. Oxford University Press, Oxford/New York/Toronto, 1993.

[Fed94] K. Fedra. Model-based environmental information and decision support systems. In L. M. Hilty, A. Jaeschke, B. Page, and A. Schwabl, editors, *Informatik für den Umweltschutz*. Metropolis, Marburg, Germany, 1994. Vol. 1.

[FG96] C. Faloutsos and V. Gaede. Analysis of $n$-dimensional quadtrees using the Hausdorff fractal dimension. In *Proc. 22nd Int. Conf. on Very Large Data Bases*, 1996.

[FK86] A. Frank and W. Kuhn. Cell graph: A provable correct method for the storage of geometry. In D. Marble, editor, *Proc. Second Int. Symp. on Spatial Data Handling*, Seattle, Wash., 1986. International Geographic Union.

[FK94] C. Faloutsos and I. Kamel. Beyond uniformity and independence: Analysis of R-trees using the concept of fractal dimension. In *Proc. 13th ACM SIGACT-SIGMOD-SIGART Symp. on Principles of Database Systems*, pages 4–13, 1994.

[FKN80] H. Fuchs, Z. Kedem, and B. Naylor. On visible surface generation by a priori tree structures. *Computer Graphics*, 14(3), 1980.

[FNPS79] R. Fagin, J. Nievergelt, N. Pippenger, and R. Strong. Extendible hashing: A fast access method for dynamic files. *ACM Trans. Database Systems*, 4(3):315–344, 1979.

[FR89] C. Faloutsos and S. Roseman. Fractals for secondary key retrieval. In *Proc. 8th ACM SIGACT/SIGMOD Symp. on Principles of Database Systems*, pages 247–252, 1989.

[FR91] C. Faloutsos and Y. Rong. DOT: a spatial access method using fractals. In *Proc. 7th IEEE Int. Conf. on Data Eng.*, pages 152–159, 1991.

[Fra82] A. Frank. Mapquery – database query language for retrieval of geometric data and its graphical representation. *ACM Computer Graphics*, 16:199–207, 1982.

[Fre87] M. Freeston. The BANG file: A new kind of grid file. In *Proc. ACM SIGMOD Int. Conf. on Management of Data*, pages 260–269, 1987.

[Fre90a] M. Freeston. Advances in the design of the BANG file. In *Proc. 3rd Int. Conf. on Foundations of Data Organization and Algorithms*. LNCS 367, pages 322–338. Springer-Verlag, Berlin/Heidelberg/New York, 1990.

[Fre90b] M. Freeston. A well-behaved structure for the storage of geometric objects. In A. Buchmann, O. Günther, T. R. Smith, and Y.-F. Wang, editors, *Design and Implementation of Large Spatial Databases*. LNCS 409, pages 287–300. Springer-Verlag, Berlin/Heidelberg/New York, 1990.

[Fre94] J. Frew. Bigfoot: An earth science computing environment for the Sequoia 2000 project. In W. K. Michener, J. W. Brunt, and S. G. Stafford, editors, *Environmental Information Management and Analysis: Ecosystem to Global Scales*. Taylor & Francis, London, 1994.

[Fre95] M. Freeston. A general solution of the $n$-dimensional B-tree problem. In *Proc. ACM SIGMOD Int. Conf. on Management of Data*, pages 80–91, 1995.

[Fri92] J.-S. Fritz. Environmental monitoring and information management programmes of international organizations. Technical report, United Nations Environmental Program, Harmonization of Environmental Measurement (HEM) Office, GSF Neuherberg, Germany, 1992.

[FSR87] C. Faloutsos, T. Sellis, and N. Roussopoulos. Analysis of object-oriented spatial access methods. In *Proc. ACM SIGMOD Int. Conf. on Management of Data*, 1987.

[Gae95a] V. Gaede. Geometric information makes spatial query processing more efficient. In *Proc. 3rd ACM International Workshop on Advances in Geographic Information Systems (ACM-GIS'95)*, pages 45–52, Baltimore, Md., 1995.

[Gae95b] V. Gaede. Optimal redundancy in spatial database systems. In M. J. Egenhofer and J. R. Herring, editors, *Advances in Spatial Databases*. LNCS 951, pages 132–151. Springer-Verlag, Berlin/Heidelberg/New York, 1995.

[Gaf96] J. Gaffney. Illustra's Web DataBlade module. *SIGMOD Record*, 25(1):105–112, 1996.

[Gar97] K. Gardels. Open GIS and on-line environmental libraries. *SIGMOD Record*, 26(1):32–38, 1997.

[Gay91] P. Gayle Alston. Environment online: The greening of databases. Part 2: Scientific and technical databases. *Database*, 14(5):34–52, 1991.

[GB90] O. Günther and A. Buchmann. Research issues in spatial databases. *SIGMOD Record*, 19(4):61–68, 1990.

[GB91] O. Günther and J. Bilmes. Tree-based access methods for spatial databases: Implementation and performance evaluation. *IEEE Trans. Knowledge and Data Eng.*, 3(3):342–356, 1991.

[Geo95] Geographic Designs Inc. Online documentation of Geolineus 3.0. Santa Barbara, Calif., 1995.

[GG95] V. Gaede and O. Günther. Efficient processing of queries containing user-defined functions. In *Proc. 4th Int. Conf. on Deductive and Object-Oriented Databases*. LNCS 1013, pages 281–298. Springer-Verlag, Berlin/Heidelberg/New York, 1995.

[GG96] O. Günther and V. Gaede. Oversize shelves: A storage management technique for large spatial data objects. *Int. J. Geographical Information Systems*, 10(8), 1996.

[GG98] V. Gaede and O. Günther. Multidimensional access methods. *ACM Comp. Surv.*, 30, 1998.

[GHM+93] O. Günther, G. Hess, M. Mutz, W.-F. Riekert, and T. Ruwwe. RESEDA: a knowledge-based advisory system for remote sensing. *Applied Intelligence*, 3(4), 1993.

[GHP95] R. Grützner, A. Häuslein, and B. Page. Werkzeuge für die Umweltmodellierung und -simulation. In B. Page and L. M. Hilty, editors, *Umweltinformatik – Informatikmethoden für Umweltschutz und Umweltforschung*. Oldenbourg, Munich/Vienna, 2nd edition, 1995.

[GJR+97] W. Geiger, A. Jaeschke, O. Rentz, E. Simon, Th. Spengler, L. Zilliox, and T. Zundel, editors. *Umweltinformatik '97 – Informatique pour l'Environnement '97*. Metropolis, Marburg, Germany, 1997. Two volumes.

[GL94] O. Günther and J. Lamberts. Object-oriented techniques for the management of geographic and environmental data. *The Computer J.*, 37(1):16–25, 1994.

[GLS96] O. Günther, H. Lessing, and W. Swoboda. UDK: A European environmental data catalogue. In *Proc. Third International Conf. on Integrating GIS and Environmental Modeling*, 1996. URL http://www.ncgia.ucsb.edu/conf/santa_fe.html.

[GM88] E. D. Gilles and W. Marquardt. Prozesssimulation – ein Beitrag zum aktiven Umweltschutz. In *Beitrag der Mikroelektronik zum Umweltschutz*, GME-Fachbericht, Berlin, 1988. VDE-Verlag.

[GMS+97] O. Günther, R. Müller, P. Schmidt, H. K. Bhargava, and R. Krishnan. MMM: A WWW-based approach for sharing statistical software modules. *IEEE Internet Computing*, 1(3), 1997.

[GN91] O. Günther and H. Noltemeier. Spatial database indices for large extended objects. In *Proc. 7th IEEE Int. Conf. on Data Eng.*, 1991.

[Goo87] D. G. Goodenough et al. An expert system for remote sensing. *IEEE Trans. Geosci. Remote Sensing*, 25(3):349–359, 1987.

[GOP+97] O. Günther, V. Oria, P. Picouet, J.-M. Saglio, and M. Scholl. Benchmarking spatial joins à la carte. In J. Ferrieé, editor, *Proc. 13e Journées Bases de Données Avancées*, Grenoble, 1997.

[Gos94] J. R. Gosz. Sustainable biosphere initiative: Data management challenges. In W. K. Michener, J. W. Brunt, and S. G. Stafford, editors, *Environmental Information Management and Analysis: Ecosystem to Global Scales*, chapter 3, pages 27–39. Taylor & Francis, London, 1994.

[GP94] G. Guariso and B. Page, editors. *Computer Support for Environmental Impact Assessment*, volume B-16 of *IFIP Transactions*. North-Holland, Amsterdam, 1994.

[GPS93] M. F. Goodchild, B. O. Parks, and L. T. Steyaert, editors. *Environmental Modeling with GIS*. Oxford University Press, New York/Oxford, 1993.

[GR92] O. Günther and W.-F. Riekert. *Wissensbasierte Methoden zur Fernerkundung der Umwelt*. Wichmann, Karlsruhe, 1992.

[GR93] O. Günther and W.-F. Riekert. The design of GODOT: An object-oriented geographic information system. *IEEE Data Engineering Bulletin*, 16(3), 1993.

[GR94] V. Gaede and W.-F. Riekert. Spatial access methods and query processing in the object-oriented GIS GODOT. In *Proc. AGDM'94 Workshop*, pages 40–52, Delft, The Netherlands, 1994. Netherlands Geodetic Commission.

[Gre89] D. Greene. An implementation and performance analysis of spatial data access methods. In *Proc. 5th IEEE Int. Conf. on Data Eng.*, 1989.

[GRR95] O. Günther, F. J. Radermacher, and W.-F. Riekert. Environmental monitoring: Models, methods, and systems. In N. M. Avouris and B. Page, editors, *Environmental Informatics – Methodology and Applications of Environmental Information Processing*, chapter 3, pages 13–38. Kluwer Academic Publishers, Norwell, Mass., 1995.

[Grü97] R. Grützner, editor. *Modellierung und Simulation im Umweltbereich*. Vieweg, Braunschweig/Wiesbaden, 1997.

[GS91] O. Günther and H.-J. Schek, editors. *Advances in Spatial Databases*. LNCS 525. Springer-Verlag, Berlin/Heidelberg/New York, 1991.

[GS92] P. Gayle Alston and F. W. Stoss. Environment online: The greening of databases. Part 3: Business and regulatory information. *Database*, 14(6):17–35, 1992.

[GS93] R. H. Güting and M. Schneider. Realms: A foundation for spatial data types in database systems. In D. Abel and B. C. Ooi, editors, *Advances in Spatial Databases*. LNCS 692. Springer-Verlag, Berlin/Heidelberg/New York, 1993.

[GS95] R. H. Güting and M. Schneider. Realm-based spatial data types: The ROSE algebra. *The VLDB J.*, 4:100–143, 1995.

[GSP96] M. F. Goodchild, L. T. Steyaert, and B. O. Parks, editors. *GIS and Environmental Modeling: Progress and Research Issues*. GIS World Books, 1996.

[Gün88] O. Günther. *Efficient Structures for Geometric Data Management*. LNCS 337. Springer-Verlag, Berlin/Heidelberg/New York, 1988.

[Gün89] O. Günther. The cell tree: An object-oriented index structure for geometric databases. In *Proc. 5th IEEE Int. Conf. on Data Eng.*, 1989.

[Gün92] O. Günther. Evaluation of spatial access methods with oversize shelves. In G. Gambosi, M. Scholl, and H.-W. Six, editors, *Geographic Database Management Systems*. Springer-Verlag, Berlin/Heidelberg/New York, 1992.

[Gün93] O. Günther. Efficient computation of spatial joins. In *Proc. 9th IEEE Int. Conf. on Data Eng.*, 1993.

[Gün97] O. Günther, editor. Special section on environmental information systems. *SIGMOD Record*, 26(1):3–38, 1997.

[Gut84] A. Guttman. R-trees: A dynamic index structure for spatial searching. In *Proc. ACM SIGMOD Int. Conf. on Management of Data*, pages 47–54, 1984.

[Güt89] R. H. Güting. Gral: An extendible relational database system for geometric applications. In *Proc. 15th Int. Conf. on Very Large Data Bases*, pages 33–44, 1989.

[Güt94] R. H. Güting. An introduction to spatial database systems. *The VLDB J.*, 3(4):357–399, 1994.

[GV98] O. Günther and A. Voisard. Metadata in Geographic and Environmental Data Management. In W. Klas and A. Sheth, editors, *Managing Multimedia Data: Using Metadata to Integrate and Apply Digital Data*. McGraw-Hill, New York, 1998.

[GW89a] G. Guariso and H. Werthner, editors. *Environmental Decision Support Systems*. John Wiley & Sons, New York, 1989.

[GW89b] O. Günther and E. Wong. Convex polyhedral chains: A representation for geometric data. *Computer-Aided Design*, 21(3), 1989.

[GW91]  O. Günther and E. Wong. A dual approach to detect polyhedral intersections in arbitrary dimensions. *BIT*, 31:2–14, 1991.

[GWK88]  M. W. Gery, G. Z. Whitten, and J. P. Killus. Development and testing of the CBM-IV for urban and regional modeling. Technical Report EPA-600/3-88-012, United States Environmental Protection Agency, 1988.

[Hal95]  G. Hallmark. The Oracle Warehouse. In *Proc. 21st Int. Conf. on Very Large Data Bases*, 1995.

[Han92]  P. Hane. *Environment Online: The Greening of Databases. The Complete Environmental Series from Database Magazine*. Eight Bit Books, Wilton, UK, 1992.

[HCL⁺90]  L. Haas, W. Chang, G.M. Lohman, J. McPherson, P.F. Wilms, G. Lapis, B. Lindsay, H. Pirahesh, M. J. Carey, and E. Shekita. Starburst mid-flight: As the dust clears. *IEEE Trans. Knowledge and Data Eng.*, 2(1):143–161, 1990.

[Hen93]  I. Henning. Von Sachdaten zu Führungsinformation. In A. Jaeschke, T. Kämpke, B. Page, and F. J. Radermacher, editors, *Informatik im Umweltschutz*. Informatik aktuell. Springer-Verlag, Berlin/Heidelberg/New York, 1993.

[HG96]  T. Holland and O. Günther. Integration von Umweltaspekten in die betriebliche Informationsverarbeitung am Beispiel von SAP R/3. In H. Lessing and U. Lipeck, editors, *Informatik im Umweltschutz*. Metropolis, Marburg, Germany, 1996.

[HHH⁺95]  H.-D. Haasis, L. M. Hilty, J. Hunscheid, H. Kürzl, and C. Rautenstrauch, editors. *Umweltinformationssysteme in der Produktion*. Metropolis, Marburg, Germany, 1995.

[HHKR95]  H.-D. Haasis, L. M. Hilty, H. Kürzl, and C. Rautenstrauch, editors. *Betriebliche Umweltinformationssysteme – Projekte und Perspektiven*. Metropolis, Marburg, Germany, 1995.

[Hil95]  L. M. Hilty. Information systems for industrial environmental management. In N. M. Avouris and B. Page, editors, *Environmental Informatics – Methodology and Applications of Environmental Information Processing*, chapter 22, pages 371–384. Kluwer Academic Publishers, Norwell, Mass., 1995.

[Hin85]  K. Hinrichs. Implementation of the grid file: Design concepts and experience. *BIT*, 25:569–592, 1985.

[HJPS94]  L. M. Hilty, A. Jaeschke, B. Page, and A. Schwabl, editors. *Informatik für den Umweltschutz*. Metropolis, Marburg, Germany, 1994. Two volumes.

[HKT95]  W. Härdle, S. Klinke, and B. A. Turlach. *XploRe: An Interactive Statistical Computing Environment*. Springer-Verlag, Berlin/Heidelberg/New York, 1995.

[HLS88]  J. Herring, R. Larsen, and J. Shivakumar. Extensions to the SQL language to support spatial analysis in a topological data base. In *Proc. GIS/LIS Conf.*, pages 741–750, San Antonio, Tex., 1988.

[HMW96]  R. Haining, J. Ma, and S. Wise. Design of a software system for interactive spatial statistical analysis linked to a GIS. *Computational Statistics*, 11(4):449–466, 1996.

[HNP95]  J. M. Hellerstein, J. F. Naughton, and A. Pfeffer. Generalized search trees for database systems. In *Proc. 21st Int. Conf. on Very Large Data Bases*, pages 562–573, 1995.

[HPRR95]  L. M. Hilty, B. Page, F. J. Radermacher, and W.-F. Riekert. Environmental informatics as a new discipline of applied computer science. In N. M. Avouris and B. Page, editors, *Environmental Informatics – Methodology and Applications of Environmental Information Processing*, chapter 1, pages 1–11. Kluwer Academic Publishers, Norwell, Mass., 1995.

[HR85] T. Härder and A. Reuter. Architecture of database systems for non-standard applications. In A. Blaser and P. Pistor, editors, *Datenbanksysteme in Büro, Technik und Wissenschaft.* Informatik-Fachberichte 94, pages 253–286. Springer-Verlag, Berlin/Heidelberg/New York, 1985.

[HR94] H.-D. Haasis and O. Rentz. PPS-Systeme zur Unterstützung des betrieblichen Umweltschutzes. *CIM Management,* 10(3):48–53, 1994.

[HR95] L. M. Hilty and C. Rautenstrauch. Betriebliche Umweltinformatik. In B. Page and L. M. Hilty, editors, *Umweltinformatik – Informatikmethoden für Umweltschutz und Umweltforschung.* Oldenbourg, Munich/Vienna, 2nd edition, 1995.

[HR97] L. M. Hilty and C. Rautenstrauch. Betriebliche Umweltinformationssysteme (BUIS) – eine Literaturanalyse. *Informatik-Spektrum,* 20(3), 1997.

[HS92a] A. Henrich and H.-W. Six. How to split buckets in spatial data structures. In G. Gambosi, M. Scholl, and H.-W. Six, editors, *Geographic Database Management Systems,* pages 212–244. Springer-Verlag, Berlin/Heidelberg/New York, 1992.

[HS92b] E. G. Hoel and H. Samet. A qualitative comparison study of data structures for large segment databases. In *Proc. ACM SIGMOD Int. Conf. on Management of Data,* pages 205–214, 1992.

[HS95] E. G. Hoel and H. Samet. Benchmarking spatial join operations with spatial output. In *Proc. 21st Int. Conf. on Very Large Data Bases,* pages 606–618, 1995.

[HSW88a] A. Hutflesz, H.-W. Six, and P. Widmayer. Globally order preserving multidimensional linear hashing. In *Proc. 4th IEEE Int. Conf. on Data Eng.,* pages 572–579, 1988.

[HSW88b] A. Hutflesz, H.-W. Six, and P. Widmayer. Twin grid files: Space optimizing access schemes. In *Proc. ACM SIGMOD Int. Conf. on Management of Data,* 1988.

[HSW89] A. Henrich, H.-W. Six, and P. Widmayer. The LSD tree: Spatial access to multidimensional point and non-point objects. In *Proc. 15th Int. Conf. on Very Large Data Bases,* pages 45–53, 1989.

[HSW90] A. Hutflesz, H.-W. Six, and P. Widmayer. The R-file: An efficient access structure for proximity queries. In *Proc. 6th IEEE Int. Conf. on Data Eng.,* pages 372–379, 1990.

[Hüb92] M. Hübner. EXCEPT – Ein System zur Unterstützung und Dokumentation von Bewertungsvorgängen in der Umweltverträglichkeitsprüfung. In O. Günther, H. Kuhn, R. Mayer-Föll, and F. J. Radermacher, editors, *Konzeption und Einsatz von Umweltinformationssystemen.* Informatik-Fachberichte 301. Springer-Verlag, Berlin/Heidelberg/New York, 1992.

[HWS97] I. Henning, G. Wiest, and F. Schmidt. UFIS II im Rahmen des UIS Baden-Württemberg – Informationsrecherche auf neuen Wegen. In W. Geiger, A. Jaeschke, O. Rentz, E. Simon, Th. Spengler, L. Zilliox, and T. Zundel, editors, *Umweltinformatik '97 – Informatique pour l'Environnement '97.* Metropolis, Marburg, Germany, 1997. Vol. 1.

[HWZ92] A. Hutflesz, P. Widmayer, and C. Zimmermann. Global order makes spatial access faster. In G. Gambosi, M. Scholl, and H.-W. Six, editors, *Geographic Database Management Systems,* pages 161–176. Springer-Verlag, Berlin/Heidelberg/New York, 1992.

[Inf97] Informix Inc. *Informix Spatial DataBlade Module – User's Guide (Version 2.2).* Menlo Park, Calif., 1997. Part No. 000-3713.

[IP87] K. Ingram and W. Phillips. Geographic information processing using a SQL-based query language. In N. Chrisman, editor, *Proc. AUTO-CARTO 8,* pages 326–335, Baltimore, Md., 1987.

[Jag90a] H. V. Jagadish. Linear clustering of objects with multiple attributes. In *Proc. ACM SIGMOD Int. Conf. on Management of Data*, pages 332–342, 1990.

[Jag90b] H. V. Jagadish. Spatial search with polyhedra. In *Proc. 6th IEEE Int. Conf. on Data Eng.*, pages 311–319, 1990.

[JKPR93] A. Jaeschke, T. Kämpke, B. Page, and F. J. Radermacher, editors. *Informatik im Umweltschutz.* Informatik aktuell. Springer-Verlag, Berlin/Heidelberg/New York, 1993.

[JSTC91] D. Johnson, P. Shelley, M. Taylor, and S. Callahan. The FINDAR directory system: a meta-model for metadata. In D. Medykyj-Scott, I. Newman, C. Ruggles, and D. Walker, editors, *Metadata in the Geosciences*, pages 123–137, Loughborough, UK, 1991.

[KBS91] H. P. Kriegel, T. Brinkhoff, and R. Schneider. An efficient map overlay algorithm based on spatial access methods and computational geometry. In G. Gambosi, M. Scholl, and H.-W. Six, editors, *Geographic Database Management Systems*, pages 194–211. Springer-Verlag, Berlin/Heidelberg/New York, 1991.

[Kel95] H. B. Keller. Neural nets in environmental applications. In N. M. Avouris and B. Page, editors, *Environmental Informatics – Methodology and Applications of Environmental Information Processing*, chapter 9, pages 127–146. Kluwer Academic Publishers, Norwell, Mass., 1995.

[KF92] I. Kamel and C. Faloutsos. Parallel R-trees. In *Proc. ACM SIGMOD Int. Conf. on Management of Data*, pages 195–204, 1992.

[KF93] I. Kamel and C. Faloutsos. On packing R-trees. In *Proc. 2nd Int. Conf. on Information and Knowledge Management*, pages 490–499, 1993.

[KF94] I. Kamel and C. Faloutsos. Hilbert R-tree: An improved R-tree using fractals. In *Proc. 20th Int. Conf. on Very Large Data Bases*, pages 500–509, 1994.

[KHH+92] H.-P. Kriegel, P. Heep, S. Heep, M. Schiwietz, and R. Schneider. An access method based query processor for spatial database systems. In G. Gambosi, M. Scholl, and H.-W. Six, editors, *Geographic Database Management Systems*, pages 273–292. Springer-Verlag, Berlin/Heidelberg/New York, 1992.

[KKN+96] A. Koschel, R. Kramer, R. Nikolai, W. Hagg, and J. Wiesel. A federation architecture for an environmental information system incorporating GIS, the World Wide Web, and CORBA. In *Proc. Third International Conf. on Integrating GIS and Environmental Modeling*, 1996. URL http://www.ncgia.ucsb.edu/conf/santa_fe.html.

[KKT+96] A. Koschel, R. Kramer, D. Theobald, G. von Bültzingsloewen, W. Hagg, J. Wiesel, and M. Müller. Evaluierung und Einsatzbeispiele von CORBA-Implementierungen für Umweltinformationssysteme. In H. Lessing and U. Lipeck, editors, *Informatik im Umweltschutz.* Metropolis, Marburg, Germany, 1996.

[KMB91] H. Keune, A. B. Murray, and H. Benking. Harmonization of environmental measurement. *GeoJournal*, 23(3):249–255, 1991.

[KNK+97] R. Kramer, R. Nikolai, A. Koschel, C. Rolker, and P. Lockemann. WWW-UDK: A Web-based environmental meta-information system. *SIGMOD Record*, 26(1):16–21, 1997.

[Koh93] J. Kohm. Das Technosphäre- und Luft-Informationssystem als übergreifende Komponente des Umweltinformationssystems Baden-Württemberg. In A. Jaeschke, T. Kämpke, B. Page, and F. J. Radermacher, editors, *Informatik im Umweltschutz.* Informatik aktuell. Springer-Verlag, Berlin/Heidelberg/New York, 1993.

[KP95] H. Kremers and W. Pillmann, editors. *Raum und Zeit in Umweltinformationssystemen.* Metropolis, Marburg, Germany, 1995. Two volumes.

[KQ96] R. Kramer and T. Quellenberg. Global access to environmental informa-
tion. In R. Denzer, G. Schimak, and D. Russell, editors, *Proc. 1995 International
Symposium on Environmental Software Systems*, pages 209–218. Chapman and
Hall, New York, 1996.

[KR95] K. Kurbel and C. Rautenstrauch. Integrated planning of production and
recycling processes: An MRP II based approach. In *Proc. 2nd Int. Conf. on
Managing Integrated Manufacturing*, pages 189–194. Management Centre, University of Leicester, Leicester, UK, 1995.

[Kra93] J. Kramer. Betriebliche Umweltinformationssysteme (BUIS): Vorausset-
zung effektiven Umweltmanagements. In H.-K. Arndt, editor, *Umweltinformationssysteme für Unternehmen*, volume 69/93 of *Schriftenreihe des IÖW*.
Institut für ökologische Wirtschaftsforschung (IÖW), Berlin, 1993.

[Kri84] H.-P. Kriegel. Performance comparison of index structures for multikey
retrieval. In *Proc. ACM SIGMOD Int. Conf. on Management of Data*, pages
186–196, 1984.

[KS86] H.-P. Kriegel and B. Seeger. Multidimensional order preserving linear hashing with partial expansions. In *Proc. Int. Conf. on Database Theory*. LNCS 243.
Springer-Verlag, Berlin/Heidelberg/New York, 1986.

[KS87] H.-P. Kriegel and B. Seeger. Multidimensional quantile hashing is very
efficient for non-uniform record distributions. In *Proc. 3rd IEEE Int. Conf. on
Data Eng.*, pages 10–17, 1987.

[KS88] H.-P. Kriegel and B. Seeger. PLOP-hashing: A grid file without directory.
In *Proc. 4th IEEE Int. Conf. on Data Eng.*, pages 369–376, 1988.

[KS89] H.-P. Kriegel and B. Seeger. Multidimensional quantile hashing is very
efficient for non-uniform distributions. *Information Sciences*, 48:99–117, 1989.

[KS91] C. Kolovson and M. Stonebraker. Segment indexes: Dynamic indexing techniques for multi-dimensional interval data. In *Proc. ACM SIGMOD Int. Conf.
on Management of Data*, pages 138–147, 1991.

[KSE95] K. Kurbel, B. Schneider, and A. Etzrodt. Von Produktionsdaten zu Re-
cyclinginformationen. In H.-D. Haasis, L. M. Hilty, J. Hunscheid, H. Kürzl,
and C. Rautenstrauch, editors, *Umweltinformationssysteme in der Produktion*.
Metropolis, Marburg, Germany, 1995.

[KSSS90] H.-P. Kriegel, M. Schiwietz, R. Schneider, and B. Seeger. Performance comparison of point and spatial access methods. In A. Buchmann,
O. Günther, T. R. Smith, and Y.-F. Wang, editors, *Design and Implementation of Large Spatial Databases*. LNCS 409, pages 89–114. Springer-Verlag,
Berlin/Heidelberg/New York, 1990.

[Kug95] R. Kuggeleijn. Managing data about data. *GIS Europe*, 4(3):32–33, 1995.

[Kum94] A. Kumar. A study of spatial clustering techniques. In D. Karagiannis, editor, *Proc. 5th Conf. on Database and Expert Systems Applications (DEXA'94)*.
LNCS 856, pages 57–70. Springer-Verlag, Berlin/Heidelberg/New York, 1994.

[KW87] A. Kemper and M. Wallrath. An analysis of geometric modeling in
database systems. *ACM Comp. Surv.*, 19(1):47–91, 1987.

[Lad97] P. Ladstätter. Geodaten-Management mit SICAD/GDB-X: Neue Entwick-
lungen und Positionierung. In *Proc. 5th Int. User Forum*, Munich, 1997. Siemens
Nixdorf.

[Lan92] G. Langran. *Time in Geographic Information Systems*. Taylor & Francis,
London, 1992.

[Lar80] P. A. Larson. Linear hashing with partial expansions. In *Proc. 6th Int.
Conf. on Very Large Data Bases*, pages 224–232, 1980.

[LCF+91] R. Lunetta, R. Congalton, L. Fenstermaker, J. Jensen, K. McTwire, and
L. Tinney. Remote sensing and geographic information system data integration:

Error sources and research issues. *Photogrammetric Engineering and Remote Sensing*, 57(6):677–687, 1991.

[Les89]  H. Lessing. Umweltinformationssysteme – Anforderungen und Möglichkeiten am Beispiel Niedersachsens. In A. Jaeschke, W. Geiger, and B. Page, editors, *Informatik im Umweltschutz*. Springer-Verlag, Berlin/Heidelberg/New York, 1989.

[Lit80]  W. Litwin. Linear hashing: A new tool for file and table addressing. In *Proc. 6th Int. Conf. on Very Large Data Bases*, pages 212–223, 1980.

[LK87]  T. M. Lillesand and R. W. Kiefer. *Remote Sensing and Image Interpretation*. John Wiley & Sons, New York, 1987.

[LKH+94]  R. Lenz, M. Knorrenschild, C. Herderich, O. Springstobbe, E. Forster, J. Benz, W. Assoff, and W. Windhorst. An information system of ecological models. Technical Report 27/94, GSF, Oberschleissheim, Germany, 1994. URL http://www.gsf.de/UFIS/ufis.

[LL96]  H. Lessing and U. Lipeck, editors. *Informatik im Umweltschutz*. Metropolis, Marburg, Germany, 1996.

[LLOW91]  C. Lamb, G. Landis, J. Orenstein, and D. Weinreb. The ObjectStore database system. *Comm. ACM*, 10(34):50–63, 1991.

[LLPS91]  G. Lohman, B. Lindsay, H. Pirahesh, and K. B. Schiefer. Extensions to Starburst: Objects, types, functions and rules. *Comm. ACM*, 34(10):94–109, 1991.

[LM88]  R. Laurini and F. Milleret. Spatial data base queries: Relational algebra versus computational geometry. In M. Rafanelli, J. Klensin, and P. Svensson, editors, *Proc. 4th Int. Conf. on Statistical and Scientific Database Management*. LNCS 339, pages 291–313. Springer-Verlag, Berlin/Heidelberg/New York, 1988.

[LMOB91]  T. R. Loveland, J. M. Merchant, D. O. Ohlen, and J. F. Brown. Development of a land-cover characteristics database for the conterminous U.S. *Photogrammetric Engineering and Remote Sensing*, 57(11):1453–1463, 1991.

[LO93]  T. R. Loveland and D. O. Ohlen. Experimental AVHRR land data sets for environmental monitoring and modeling. In M. F. Goodchild, B. O. Parks, and L. T. Steyaert, editors, *Environmental Modeling with GIS*, chapter 37, pages 379–385. Oxford University Press, New York/Oxford, 1993.

[LPS93]  A. Lott, M. Pauleser, and W. Strauss. Umwelt- und Arbeitsschutzmanagementsysteme für den Einsatz in Industrie und Gewerbe. In H.-K. Arndt, editor, *Umweltinformationssysteme für Unternehmen*, volume 69/93 of *Schriftenreihe des IÖW*. Institut für ökologische Wirtschaftsforschung (IÖW), Berlin, 1993.

[LR94]  M. L. Lo and C. V. Ravishankar. Spatial joins using seeded trees. In *Proc. ACM SIGMOD Int. Conf. on Management of Data*, pages 209–220, 1994.

[LS89a]  D. B. Lomet and B. Salzberg. The hB-tree: A robust multiattribute search structure. In *Proc. 5th IEEE Int. Conf. on Data Eng.*, pages 296–304, 1989.

[LS89b]  G. F. Luger and W. A. Stubblefield. *Artificial Intelligence and the Design of Expert Systems*. Benjamin/Cummings, Redwood City, Calif., 1989.

[LS90]  D. B. Lomet and B. Salzberg. The hB-tree: A multiattribute indexing method with good guaranteed performance. *ACM Trans. Database Systems*, 15(4):625–658, 1990. Reprinted in [Sto94].

[LS94]  H. Lessing and T. Schütz. Der Umwelt-Datenkatalog als Instrument zur Steuerung von Informationsflüssen. In L. M. Hilty, A. Jaeschke, B. Page, and A. Schwabl, editors, *Informatik für den Umweltschutz*. Metropolis, Marburg, Germany, 1994.

[LT94]  R. Laurini and D. Thompson. *Fundamentals of Spatial Information Systems*. Academic Press, London/San Diego, 1994.

[Mat96]  MathSoft Inc. *S+Gislink*. Seattle, Wash., 1996.

[May93] R. Mayer-Föll. Das Umweltinformationssystem Baden-Württemberg. In A. Jaeschke, T. Kämpke, B. Page, and F. J. Radermacher, editors, *Informatik im Umweltschutz*. Informatik aktuell. Springer-Verlag, Berlin/Heidelberg/New York, 1993.

[May97] R. Mayer-Föll. Umweltinformationssystem Baden-Württemberg. In *Proc. InterGeo'97*, Karlsruhe, Germany, 1997.

[MB94] D. Miller and K. Bullock. Metadata for land and geographic information – an Australia-wide framework. In *Proc. AURISA'94*, pages 391–398, Sydney, 1994.

[MBS94] W. K. Michener, J. W. Brunt, and S. G. Stafford, editors. *Environmental Information Management and Analysis: Ecosystem to Global Scales*. Taylor & Francis, London, 1994.

[McC82] J. McCarthy. Metadata management for large statistical databases. In *Proc. 8th Int. Conf. on Very Large Data Bases*, 1982.

[McC94] Rob McCool. *The Common Gateway Interface*. http://hoo-hoo.ncsa.uiuc.edu/cgi/overview.html, 1994.

[McK87] D. M. McKeown. The role of artificial intelligence in the integration of remotely sensed data with geographic information systems. *IEEE Trans. Geosci. Remote Sensing*, 25(3):330–347, 1987.

[McM87] R. B. McMaster. Automated line generalization. *Cartographica*, 24(2):74–111, 1987.

[McM88] R. B. McMaster. Cartographic generalization in a digital enviroment: A framework for implementation in geographic information systems. In *Proc. GIS/LIS Conf.*, 1988.

[MDF95] R. B. Melton, D. M. DeVaney, and J. C. French, editors. *The Role of metadata in managing large environmental science datasets (Proc. SDM-92)*, Richland, Wash., 1995. Pacific Northwest Laboratory. Technical Report No. PNL-SA-26092.

[MEC94] G. G. Moisen, T. C. Edwards, and D. R. Cutler. Spatial sampling to assess classification accuracy of remotely sensed data. In W. K. Michener, J. W. Brunt, and S. G. Stafford, editors, *Environmental Information Management and Analysis: Ecosystem to Global Scales*, chapter 11, pages 159–176. Taylor & Francis, London, 1994.

[MF94] D. Miller and B. Forner. Experience in developing a natural resource data directory for New South Wales. In *Proc. AURISA'94*, pages 391–398, Sydney, 1994.

[MGR91] D. J. Maguire, M. F. Goodchild, and D. W. Rhind, editors. *Geographical Information Systems: Principles and Applications*. Longman, Harlow, UK, 1991. Two volumes.

[Mie93] P. Miettinen. Software tools in life cycle assessment. In B. Pedersen Weidema, editor, *Environmental Assessment of Products*. UETP-EEE – The Finnish Association of Graduate Engineers, Helsinki, 1993.

[MMS+96] E. Mesrobian, R. Muntz, E. Shek, S. Nittel, M. LaRouche, and M. Kriguer. OASIS: An open architecture scientific information system. In *Proc. RIDE'96*, 1996.

[Mor85] S. Morehouse. ARC/INFO: A geo-relational model for spatial information. In *Proc. 7th Int. Symp. on Computer Assisted Cartography*, 1985.

[MP69] M. Minsky and S. Papert. *Perceptrons*. MIT Press, Cambridge, Mass., 1969.

[MP94] C. Bauzer Medeiros and F. Pires. Databases for GIS. *SIGMOD Record*, 23:107–115, 1994.

[MR95] K. Mikkonen and A. Rainio. Towards a societal GIS in Finland – ArcView application queries data from published geographical databases. In ESRI Inc., editor, *Proc. 15th ESRI User Conference*, Redlands, Calif., 1995. ESRI Inc.

[MSNRW91] D. Medykyj-Scott, I. Newman, C. Ruggles, and D. Walker, editors. *Metadata in the Geosciences.* Group D Publications, Loughborough, UK, 1991.

[MSS96] R. Mayer-Föll, J. Strohm, and A. Schultze. Umweltinformationssystem Baden-Württemberg – Überblick Rahmenkonzeption. In H. Lessing and U. Lipeck, editors, *Informatik im Umweltschutz.* Metropolis, Marburg, Germany, 1996.

[Mül93] M. Müller. Entwicklung des Räumlichen Informations- und Planungssystems (RIPS) als übergreifende Komponente des Umweltinformationssystems Baden-Württemberg. In A. Jaeschke, T. Kämpke, B. Page, and F. J. Radermacher, editors, *Informatik im Umweltschutz.* Informatik aktuell. Springer-Verlag, Berlin/Heidelberg/New York, 1993.

[MWWW93] H. Müller-Witt, M. Wiecken, and R. Winkelmann. Betriebliches Umweltmanagement und das betriebliche Umwelt-Informations-System (BUIS). In H.-K. Arndt, editor, *Umweltinformationssysteme für Unternehmen*, volume 69/93 of *Schriftenreihe des IÖW.* Institut für ökologische Wirtschaftsforschung (IÖW), Berlin, 1993.

[Nat97] National Center for Geographic Information and Analysis (NCGIA). *Core Curriculum.* University of California, Santa Barbara, 1997. Three volumes. URL http://www.ncgia.ucsb.edu/education.

[Nat98] National Center for Geographic Information and Analysis (NCGIA), 1998. URL http://www.ncgia.ucsb.edu.

[ND97] R. G. Newell and M. Doe. Discrete geometry with seamless topology in a GIS, 1997. URL http://www.smallworld-us.com.

[NH87] J. Nievergelt and K. Hinrichs. Storage and access structures for geometric data bases. In S. Ghosh, Y. Kambayashi, and K. Tanaka, editors, *Proc. Int. Conf. on Foundations of Data Organization 1985.* Plenum Press, New York, 1987.

[NH94] R. T. Ng and J. Han. Efficient and effective clustering methods for spatial data mining. In *Proc. 20th Int. Conf. on Very Large Data Bases*, pages 144–154, 1994.

[NHS84] J. Nievergelt, H. Hinterberger, and K. C. Sevcik. The grid file: An adaptable, symmetric multikey file structure. *ACM Trans. Database Systems*, 9(1):38–71, 1984.

[NK93] V. Ng and T. Kameda. Concurrent accesses to R-trees. In D. Abel and B. C. Ooi, editors, *Advances in Spatial Databases.* LNCS 692, pages 142–161. Springer-Verlag, Berlin/Heidelberg/New York, 1993.

[NK94] V. Ng and T. Kameda. The R-link tree: A recoverable index structure for spatial data. In D. Karagiannis, editor, *Proc. 5th Conf. on Database and Expert Systems Applications (DEXA '94).* LNCS 856, pages 163–172. Springer-Verlag, Berlin/Heidelberg/New York, 1994.

[NS87] R. C. Nelson and H. Samet. A population analysis for hierarchical data structures. In *Proc. ACM SIGMOD Int. Conf. on Management of Data*, pages 270–277, 1987.

[Obj95a] Object Management Group. The Common Object Request Broker: Architecture and specification, version 2.0. Framingham, Mass., 1995.

[Obj95b] Object Management Group. CORBAservices: Common Object Services Specification. Framingham, Mass., 1995.

[Obj98] Object Store Inc., 1998. URL http://www.objectstore.com.

[OGC97] The Open GIS Consortium and the OGIS Project, 1997. URL http://www.opengis.org.

[OM84] J. Orenstein and T. H. Merrett. A class of data structures for associative searching. In *Proc. 3rd ACM SIGACT-SIGMOD Symp. on Principles of Database Systems*, pages 181–190, 1984.

[OM88] J. Orenstein and F. A. Manola. PROBE spatial data modeling and query processing in an image database application. *IEEE Trans. Software Eng.*, 14(5):611–629, 1988.

[OMSD89] B. C. Ooi, K. J. McDonell, and R. Sacks-Davis. Extending a DBMS for geographic applications. In *Proc. 5th IEEE Int. Conf. on Data Eng.*, 1989.

[Ooi90] B. C. Ooi. *Efficient Query Processing in Geographic Information Systems.* LNCS 471. Springer-Verlag, Berlin/Heidelberg/New York, 1990.

[Oos90] P. van Oosterom. *Reactive Data Structures for Geographic Information Systems.* PhD thesis, University of Leiden, The Netherlands, 1990.

[Ope91] S. Openshaw. Developing appropriate spatial analysis methods for GIS. In D. J. Maguire, M. F. Goodchild, and D. W. Rhind, editors, *Geographical Information Systems: Principles and Applications*, volume 1, chapter 25, pages 389–402. Longman, Harlow, UK, 1991.

[Ora94] Oracle Deutschland GmbH. Der Automobilbauer BMW ist für die Zukunft gerüstet: Der Umwelt und den Menschen zuliebe. *Oracle World*, pages 4–6, 1994.

[Ora95] Oracle Inc. Oracle 7 multidimension: Advances in relational database technology for spatial data management, March 1995. White Paper.

[Ore82] J. Orenstein. Multidimensional tries used for associative searching. *Information Processing Letters*, 14(4):150–157, 1982.

[Ore86] J. Orenstein. Spatial query processing in an object-oriented database system. In *Proc. ACM SIGMOD Int. Conf. on Management of Data*, pages 326–333, 1986.

[Ore89] J. Orenstein. Redundancy in spatial databases. In *Proc. ACM SIGMOD Int. Conf. on Management of Data*, pages 294–305, 1989.

[Ore90a] J. Orenstein. A comparison of spatial query processing techniques for native and parameter space. In *Proc. ACM SIGMOD Int. Conf. on Management of Data*, pages 343–352, 1990.

[Ore90b] J. Orenstein. An object-oriented approach to spatial data processing. In *Proc. 4th Int. Symp. on Spatial Data Handling*, Zürich, 1990.

[Ore90c] J. Orenstein. Strategies for optimizing the use of redundancy in spatial databases. In A. Buchmann, O. Günther, T. R. Smith, and Y.-F. Wang, editors, *Design and Implementation of Large Spatial Databases.* LNCS 409, pages 115–134. Springer-Verlag, Berlin/Heidelberg/New York, 1990.

[Ort95] A. Ortmaier. Online-Datenbanken zur Informationsgewinnung für Zwecke des Umweltschutzes – Eine praktische Anleitung zur sinnvollen Auswahl und effizienten Nutzung. Master's thesis, Humboldt-Universität, Berlin, 1995.

[OS83a] Y. Ohsawa and M. Sakauchi. BD-tree: A new N-dimensional data structure with efficient dynamic characteristics. In *Proc. 9th World Computer Congress, IFIP 1983*, pages 539–544, 1983.

[OS83b] M. Ouksel and P. Scheuermann. Storage mappings for multidimensional linear dynamic hashing. In *Proc. 2th ACM SIGACT-SIGMOD Symp. on Principles of Database Systems*, pages 90–105, 1983.

[OSBvK90] M. H. Overmars, M. Smid, T. de Berg, and M. J. van Kreveld. Maintaining range trees in secondary memory. Part I: Partitions. *Acta Informatica*, 27:423–452, 1990.

[Ost95] F. Ostyn. The EDRA – Fueling GIS applications with required geographical information. In ESRI Inc., editor, *Proc. 15th ESRI User Conference*, Redlands, Calif., 1995. ESRI Inc.

[Oto85] E. J. Otoo. Symmetric dynamic index maintenance scheme. In *Proc. Int. Conf. on Foundations of Data Organization*, pages 283–296, 1985.

[Ouk85]  M. Ouksel. The interpolation based grid file. In *Proc. 4th ACM SIGACT-SIGMOD Symp. on Principles of Database Systems*, pages 20–27, 1985.

[OV91]  P. van Oosterom and T. Vijlbrief. Building a GIS on top of the open DBMS "Postgres". In *Proc. European Conference on GIS*, pages 775–787, 1991.

[PdSMS94]  J. P. Peloux, G. Reynal de St Michel, and M. Scholl. Evaluation of spatial indices implemented with the O$_2$ DBMS. *Ingénièrie des systèmes d'information*, 6, 1994.

[PE88]  D. Pullar and M. Egenhofer. Towards formal definitions of topological relations among spatial objects. In *Proc. 3rd Int. Symp. on Spatial Data Handling*, pages 225–242, 1988.

[Ped93]  B. Pedersen Weidema, editor. *Environmental Assessment of Products*. UETP-EEE – The Finnish Association of Graduate Engineers, Helsinki, 1993.

[PH95]  B. Page and L. M. Hilty, editors. *Umweltinformatik – Informatikmethoden für Umweltschutz und Umweltforschung*. Oldenbourg, Munich/Vienna, 2nd edition, 1995.

[PK92]  W. Pillmann and D. J. Kahn. Distributed environmental data compendia. In A. M. Aiken, editor, *Proc. 12th IFIP World Computer Congress*. Elsevier, Amsterdam, 1992.

[PM90a]  D. J. Peuquet and D. F. Marble. ARC/INFO: an example of a contemporary geographic information system. In D. J. Peuquet and D. F. Marble, editors, *Introductory Readings in Geographic Information Systems*, pages 90–99. Taylor & Francis, London, 1990.

[PM90b]  D. J. Peuquet and D. F. Marble, editors. *Introductory Readings in Geographic Information Systems*. Taylor & Francis, London, 1990.

[PS85]  F. P. Preparata and M. I. Shamos. *Computational Geometry*. Springer-Verlag, Berlin/Heidelberg/New York, 1985.

[PST93]  B. U. Pagel, H.-W. Six, and H. Toben. The transformation technique for spatial objects revisited. In D. Abel and B. C. Ooi, editors, *Advances in Spatial Databases*. LNCS 692, pages 73–88. Springer-Verlag, Berlin/Heidelberg/New York, 1993.

[PSTW93]  B. U. Pagel, H.-W. Six, H. Toben, and P. Widmayer. Towards an analysis of range query performance in spatial data structures. In *Proc. 12th ACM SIGACT-SIGMOD Symp. on Principles of Database Systems*, pages 214–221, 1993.

[PSW95]  B. U. Pagel, H.-W. Six, and M. Winter. Window query optimal clustering of spatial objects. In *Proc. 14th ACM SIGACT-SIGMOD Symp. on Principles of Database Systems*, pages 86–94, 1995.

[PW95]  H.-N. Pham and T. Wittig. An adaptable architecture for river quality monitoring. In N. M. Avouris and B. Page, editors, *Environmental Informatics – Methodology and Applications of Environmental Information Processing*, chapter 14, pages 217–235. Kluwer Academic Publishers, Norwell, Mass., 1995.

[Rad91]  F. J. Radermacher. The importance of metaknowledge for environmental information systems. In O. Günther and H.-J. Schek, editors, *Advances in Spatial Databases*. LNCS 525, pages 35–44. Springer-Verlag, Berlin/Heidelberg/New York, 1991.

[Rau94]  C. Rautenstrauch. Integrating information systems for production and recycling. In L. M. Hilty, A. Jaeschke, B. Page, and A. Schwabl, editors, *Informatik für den Umweltschutz*. Metropolis, Marburg, Germany, 1994. Vol. 2.

[Reg85]  M. Regnier. Analysis of the grid file algorithms. *BIT*, 25:335–357, 1985.

[Rhy97]  T. M. Rhyne. Going virtual with geographic information and scientific visualization. *Computers and Geosciences*, 23(4), 1997.

[Ric85]  O. Richter. *Simulation des Verhaltens ökologischer Systeme – Mathematische Methoden und Modelle*. VCH-Verlag, Weinheim, Germany, 1985.

[Rie93]   W.-F. Riekert.  Extracting area objects from raster image data.  *IEEE Comp. Graphics and App.*, 13(2):68–73, 1993.

[RK91]   J. F. Raper and B. Kelk. Three-dimensional GIS. In D. J. Maguire, M. F. Goodchild, and D. W. Rhind, editors, *Geographical Information Systems: Principles and Applications*, volume 1, chapter 20, pages 299–317. Longman, Harlow, UK, 1991.

[RKV95]   G. Della Riccia, R. Kruse, and R. Viertl, editors. *Mathematical and Statistical Methods in Artificial Intelligence.* Springer-Verlag, Berlin/Heidelberg/New York, 1995.

[RL84]   N. Roussopoulos and D. Leifker.  An introduction to PSQL: A pictoral structured query language. In *IEEE Workshop on Visual Languages*, 1984.

[RL85]   N. Roussopoulos and D. Leifker. Direct spatial search on pictorial databases using packed R-trees. In *Proc. ACM SIGMOD Int. Conf. on Management of Data*, 1985.

[RMtPG85]   D. E. Rumelhart, J. L. McClelland, and the PDP Group, editors. *Parallel Distributed Processing: Explorations in the Microstructure of Cognition.* Bradford Books, Cambridge, Mass., 1985.

[RMW97]   W.-F. Riekert, R. Mayer-Föll, and G. Wiest.  Management of data and services in the Environmental Information System (UIS) of Baden-Württemberg. *SIGMOD Record*, 26(1):22–26, 1997.

[Rob81]   J. T. Robinson. The K-D-B-tree: A search structure for large multidimensional dynamic indexes. In *Proc. ACM SIGMOD Int. Conf. on Management of Data*, pages 10–18, 1981.

[Rot91]   D. Rotem.  Spatial join indices. In *Proc. 7th IEEE Int. Conf. on Data Eng.*, pages 10–18, 1991.

[RWGM97]   W.-F. Riekert, G. Wiest, M. Gaul, and B. Münst.  Management of distributed and heterogeneous information resources for environmental administrations. In R. Denzer, D. A. Swayne, and G. Schimak, editors, *Environmental Software Systems*. Chapman and Hall, London, 1997.

[Sag94]   H. Sagan.  *Space-Filling Curves.* Springer-Verlag, Berlin/Heidelberg/New York, 1994.

[Sam84]   H. Samet.  The quadtree and related hierachical data structure. *ACM Comp. Surv.*, 16(2):187–260, 1984.

[Sam90a]   H. Samet.  *Applications of Spatial Data Structures.* Addison-Wesley, Reading, Mass., 1990.

[Sam90b]   H. Samet. *The Design and Analysis of Spatial Data Structures.* Addison-Wesley, Reading, Mass., 1990.

[SAP98]   SAP AG, 1998. URL http://www.sap.com.

[SBM94]   S. G. Stafford, J. W. Brunt, and W. K. Michener. Integration of scientific information management and environmental research. In W. K. Michener, J. W. Brunt, and S. G. Stafford, editors, *Environmental Information Management and Analysis: Ecosystem to Global Scales*, chapter 1, pages 3–19. Taylor & Francis, London, 1994.

[SCB91]   D. Swayne, D. Cook, and A. Buja. XGobi: Interactive dynamic graphics in the X Window System with a link to S. In *ASA Proc. Section on Statistical Graphics*, pages 1–8. American Statistical Association, Alexandria, Va., 1991.

[SCG⁺97]   S. Shekhar, M. Coyle, B. Goyal, D.-R. Liu, and S. Sarkar. Data models in geographic information systems. *Comm. ACM*, 40(4), 1997.

[Sch91]   P. Schorn. The XYZ project: A class hierarchy and workbench for experimental geometric computation. In *Proc. Seventh Workshop on Computational Geometry.* LNCS 553. Springer-Verlag, Berlin/Heidelberg/New York, 1991.

[Sch93]   K. Scheuer. Knowledge-based interpretation of gas chromatographic data. *Chemometrics and Intelligent Laboratory Systems*, 19:201–216, 1993.

[Sch97] M. Schneider. *Spatial Data Types for Database Systems.* LNCS 1288. Springer-Verlag, Berlin/Heidelberg/New York, 1997.

[Sci97a] J&W Scientific. Sample chromatogram for herbicide detection, 1997. URL http://www.jandw.com.

[Sci97b] Scientific Computers GmbH, 1997. URL http://www.scientific.de.

[Sco94] L. M. Scott. Identification of a GIS attribute error using exploratory data analysis. *The Professional Geographer,* 46(3):378–386, 1994.

[SD78] P. H. Swain and S. M. Davis. *Remote Sensing: The Quantitative Approach.* McGraw-Hill, New York, 1978.

[SE90] J. Star and J. Estes. *Geographic Information Systems: An Introduction.* Prentice-Hall, Englewood Cliffs, N.J., 1990.

[Sea95] D. Seaborn. Database management in GIS: Is your system a poor relation? *GIS Europe,* 4(5):34–38, 1995.

[See91] B. Seeger. Performance comparison of segment access methods implemented on top of the buddy-tree. In *Advances in Spatial Databases.* LNCS 525, pages 277–296. Springer-Verlag, Berlin/Heidelberg/New York, 1991.

[SG90] T. R. Smith and P. Gao. Experimental performance evaluations on spatial access methods. In *Proc. 4th Int. Symp. on Spatial Data Handling,* Zürich, 1990.

[SGH94] M. Schmidt, J. Giegrich, and L. M. Hilty. Experiences with ecobalances and the development of an interactive software tool. In L. M. Hilty, A. Jaeschke, B. Page, and A. Schwabl, editors, *Informatik für den Umweltschutz.* Metropolis, Marburg, Germany, 1994. Vol. 2.

[Sha76] G. Shafer. *A Mathematical Theory of Evidence.* Princeton University Press, Princeton, N.J., 1976.

[SHH+96] A.-W. Scheer, H.-D. Haasis, I. Heimig, L. M. Hilty, M. Kraus, and C. Rautenstrauch, editors. *Computergestützte Stoffstrommanagement-Systeme.* Metropolis, Marburg, Germany, 1996.

[Sie98] Siemens Nixdorf Informationssysteme AG, 1998. URL http://www.sni.de.

[SK88] B. Seeger and H.-P. Kriegel. Techniques for design and implementation of spatial access methods. In *Proc. 14th Int. Conf. on Very Large Data Bases,* pages 360–371, 1988.

[SK90] B. Seeger and H.-P. Kriegel. The buddy-tree: An efficient and robust access method for spatial data base systems. In *Proc 16th Int. Conf. on Very Large Data Bases,* pages 590–601, 1990.

[SKS+97] J. Symanzik, T. Kötter, S. Schmelzer, S. Klinke, D. Cook, and D. F. Swayne. Spatial data analysis in the dynamically linked Arc-View/XGobi/XploRe environment. *Computing Science and Statistics,* 29, 1997.

[SL96] J. Seggelke and H. Lessing. Globales Umweltinformationsnetz – Eckpunkte, Chancen und Gefahren. In H. Lessing and U. Lipeck, editors, *Informatik im Umweltschutz.* Metropolis, Marburg, Germany, 1996.

[SLMS97] A. Sydow, T. Lux, P. Mieth, and R.-P. Schäfer. Simulation of traffic-induced air pollution for mesoscale applications. *Mathematics and Computers in Simulation,* 43:285–290, 1997.

[SM91] M. Siegel and S. Madnick. A metadata approach to resolving semantic conflicts. In *Proc. 17th Int. Conf. on Very Large Data Bases,* 1991.

[SMCM97] J. Symanzik, J. J. Majure, D. Cook, and I. Megretskaia. Linking Arc-View 3.0 and XGobi: Insight behind the front end. Technical Report 97-10, Department of Statistics, Iowa State University, Ames, Iowa, 1997.

[Smi96] T. R. Smith. A digital library for geographically referenced materials. *IEEE Computer,* 29(5), 1996.

[SMJ97] J. Seggelke and B. Mohaupt-Jahr. Der Verweis- und Kommunikation-sservice des Umweltbundesamtes – Ein Modellfall für das Umwelt-Intranet? In W. Geiger, A. Jaeschke, O. Rentz, E. Simon, Th. Spengler, L. Zilliox, and T. Zundel, editors, *Umweltinformatik '97 – Informatique pour l'Environnement '97*. Metropolis, Marburg, Germany, 1997. Vol. 2.

[SMN94] D. E. Strebel, B. W. Meeson, and A. K. Nelson. Scientific information systems: A conceptual framework. In W. K. Michener, J. W. Brunt, and S. G. Stafford, editors, *Environmental Information Management and Analysis: Ecosystem to Global Scales*, chapter 5, pages 59–84. Taylor & Francis, London, 1994.

[SO90] M. Smid and M. H. Overmars. Maintaining range trees in secondary memory. Part II: Lower bounds. *Acta Informatica*, 27:453–480, 1990.

[Spi93] M. Spies. *Unsicheres Wissen*. Spektrum, Heidelberg/Berlin/Oxford, 1993.

[Spo97] SpotImage. *The SPOT Satellite System*, 1997. URL http://www.spot.com.

[SPSW90] H.-J. Schek, H.-B. Paul, M. H. Scholl, and G. Weikum. The DASDBS project: Objectives, experiences and future prospects. *IEEE Trans. Knowledge and Data Eng.*, 2(1):25–43, 1990.

[SR86] M. Stonebraker and L. Rowe. The design of POSTGRES. In *Proc. ACM SIGMOD Int. Conf. on Management of Data*, 1986.

[SRF87] T. Sellis, N. Roussopoulos, and C. Faloutsos. The $R^+$-tree: A dynamic index for multi-dimensional objects. In *Proc. 13th Int. Conf. on Very Large Data Bases*, pages 507–518, 1987.

[SRH90] M. Stonebraker, L. A. Rowe, and M. Hirohama. The implementation of POSTGRES. *IEEE Trans. Knowledge and Data Eng.*, 2(1):125–142, 1990.

[SS95] M. Schmidt and A. Schorb, editors. *Stoffstromanalysen in Ökobilanzen und Öko-Audits*. Springer-Verlag, Berlin/Heidelberg/New York, 1995.

[SSH86] M. Stonebraker, T. Sellis, and E. Hanson. An analysis of rule indexing implementations in data base systems. In *Proc. 1st Int. Conf. on Expert Data Base Systems*, 1986.

[SSU91] A. Silberschatz, M. Stonebraker, and J. Ullman. Database systems: Achievements and opportunities. *Comm. ACM*, 34(10):110–120, 1991.

[SSW96] C. Schöning, R. Steinhau, and W. Wagener. LUIS – Landesumweltin-formationssystem Brandenburg: Umfassende Umweltinformationen aus erster Hand. In H. Lessing and U. Lipeck, editors, *Informatik im Umweltschutz*. Metropolis, Marburg, Germany, 1996.

[Ste93] U. Steger. *Umweltmanagement – Erfahrungen und Instrumente einer umweltorientierten Unternehmensstrategie*. Gabler, Frankfurt/Main, 1993.

[Sto90] M. Stonebraker. On rules, procedures, caching and views in data base systems. In *Proc. ACM SIGMOD Int. Conf. on Management of Data*, 1990.

[Sto91] F. W. Stoss. Environment online: The greening of databases. Part 1: General interest databases. *Database*, 14(4):13–27, 1991.

[Sto93] M. Stonebraker. The Sequoia 2000 project. In D. Abel and B. C. Ooi, editors, *Advances in Spatial Databases*. LNCS 692. Springer-Verlag, Berlin/Heidelberg/New York, 1993.

[Sto94] M. Stonebraker, editor. *Readings in Database Systems*. Morgan Kaufmann, San Mateo, 1994.

[SUN98] SUN Microsystems Inc. *JAVA Home Page*. http://java.sun.com, 1998.

[SV90] M. Scholl and A. Voisard. Thematic map modeling. In A. Buchmann, O. Günther, T. R. Smith, and Y.-F. Wang, editors, *Design and Implementation of Large Spatial Databases*. LNCS 409. Springer-Verlag, Berlin/Heidelberg/New York, 1990.

[SV92] M. Scholl and A. Voisard. Object-oriented database systems for geographic applications: An experiment with $O_2$. In F. Bancilhon, C. Delobel, and P. Kanellakis, editors, *The $O_2$ Book*, pages 585–618. Morgan Kaufmann, San Mateo, Calif., 1992.

[SV97] M. Scholl and A. Voisard, editors. *Advances in Spatial Databases*. LNCS 1262. Springer-Verlag, Berlin/Heidelberg/New York, 1997.

[SVP+96] M. Scholl, A. Voisard, J.-P. Peloux, L. Raynal, and P. Rigaux. *SGBD Géographiques*. International Thomson Publishing France, Paris, 1996.

[SW88] H. W. Six and P. Widmayer. Spatial searching in geometric databases. In *Proc. 4th IEEE Int. Conf. on Data Eng.*, pages 496–503, 1988.

[Syd94a] A. Sydow. Parallel simulation of air pollution. In *Proc. 13th IFIP World Computer Congress*. Elsevier, Amsterdam, 1994.

[Syd94b] A. Sydow. Smog analysis by parallel simulation. In L. M. Hilty, A. Jaeschke, B. Page, and A. Schwabl, editors, *Informatik für den Umweltschutz*. Metropolis, Marburg, Germany, 1994. Vol. 1.

[Syd96a] A. Sydow. Computersimulation – ein Schlüssel zum Verständnis der Umwelt. *Der GMD-Spiegel*, 26(4):16–18, 1996.

[Syd96b] A. Sydow. Modelling and simulation of air pollution. *Systems Analysis Modelling Simulation*, 25:303–314, 1996.

[Tam82] M. Tamminen. The extendible cell method for closest point problems. *BIT*, 22:27–41, 1982.

[Tam83] M. Tamminen. Performance analysis of cell based geometric file organisations. *Int. J. Comp. Vision, Graphics and Image Processing*, 24:160–181, 1983.

[The97] D. G. Theriault. Smallworld GIS: An open system architecture for a workstation-based geographical information system, 1997. URL http://www.smallworld-us.com.

[Til80] R. B. Tilove. Set Membership Classification: A Unified Approach to Geometric Intersection Problems. *IEEE Transactions on Computers*, C-29(10), 1980.

[Tom90] C. D. Tomlin. *Geographic Information Systems and Cartographic Modeling*. Prentice-Hall, Englewood Cliffs, N.J., 1990.

[TS96] Y. Theodoridis and T. K. Sellis. A model for the prediction of R-tree performance. In *Proc. 15th ACM SIGACT-SIGMOD Symp. on Principles of Database Systems*, 1996.

[TSJ91] J. Ton, J. Sticklen, and A. Jain. Knowledge-based segmentation of LANDSAT images. *IEEE Trans. Geosci. Remote Sensing*, 29(2):222–232, 1991.

[UD92] Umweltministerium Baden-Württemberg and Diebold Deutschland GmbH. *Aufbau des Informationsmanagements im UIS Baden-Württemberg – Feinkonzept für das Informationsmanagement-System (INFORMS)*. Umweltministerium Baden-Württemberg, Stuttgart, Germany, 1992.

[Ult95] A. Ultsch. Einsatzmöglichkeiten von neuronalen Netzen im Umweltbereich. In B. Page and L. M. Hilty, editors, *Umweltinformatik – Informatikmethoden für Umweltschutz und Umweltforschung*. Oldenbourg, Munich/Vienna, 1995.

[UM87] Umweltministerium Baden-Württemberg and McKinsey and Company, Inc. *Konzeption des ressortübergreifenden Umweltinformationssystems im Rahmen des Landessystemkonzepts Baden-Württemberg*. Umweltministerium Baden-Württemberg, Stuttgart, Germany, 1987. Five volumes.

[Umw92] Umweltbundesamt. Ökobilanzen für Produkte: Bedeutung – Sachstand – Perspektiven. Technical Report 38/92, Umweltbundesamt, Berlin, 1992.

[Uni92] United States Geological Survey. Spatial Data Transfer Standard (SDTS). Reston, Virg., 1992. URL http://mcmcweb.er.usgs.gov/sdts.

[Uni94a] United States Federal Geographic Data Committee. *Content Standards for Digital Geospatial Metadata.* U.S. Government, Federal Geographic Data Committee, Washington, DC, 1994. URL ftp://fgdc.er.usgs.gov.

[Uni94b] United States Government. Coordinating geographic data acquisition and access: The national spatial data infrastructure. Washington, DC, April 1994. U.S. Executive Order 12906.

[Uni97a] United States National Aeronautics and Space Administration. *Landsat Pathfinder.* U.S. Government, National Aeronautics and Space Administration,Humid Tropical Forest Inventory Project, 1997.     URL http://amazon.sr.unh.edu/pathfinder1/index.html.

[Uni97b] United States National Oceanic and Atmospheric Administration. *NESDIS Home Page.* U.S. Government, National Oceanic and Atmospheric Administration (NOAA), National Environmental Satellite, Data, and Information Service, Washington, DC, 1997. URL http://ns.noaa.gov/NESDIS/NESDIS_Home.html.

[Uni98a] University of California at Berkeley, Research Program in Environmental Planning and Geographic Information Systems.     About GRASSLinks: The public access GIS, 1998.     URL http://www.regis.berkeley.edu/grasslinks/about_gl.html.

[Uni98b] United States National Aeronautics and Space Administration. Earth observing system data and information system (EOSDIS), 1998.     URL http://spsosun.gsfc.nasa.gov/EOSDIS_main.html.

[Uni98c] University of California at Berkeley, Digital Library Project, 1998. URL http://elib.cs.berkeley.edu.

[VB93] K. Voigt and R. Brüggemann. Metadatenbank der Online Datenbanken. *Cogito*, 6:8–13, 1993.

[VB95] K. Voigt and R. Brüggemann. Meta information systems for environmental chemicals. In N. M. Avouris and B. Page, editors, *Environmental Informatics – Methodology and Applications of Environmental Information Processing*, chapter 19, pages 315–336. Kluwer Academic Publishers, Norwell, Mass., 1995.

[Voi95] A. Voisard. Mapgets: A tool for visualizing and querying geographic information. *Journal of Visual Languages and Computing*, 6:367–384, 1995.

[Vor97] F. Vorholz. Grüner Mehrwert. *Die Zeit*, page 24, 24 October 1997.

[VS94] A. Voisard and H. Schweppe. A multilayer approach to the open GIS design problem. In N. Pissinou and K. Makki, editors, *Proc. 2nd ACM GIS Workshop*, pages 23–29. ACM Press, New York, 1994.

[VS98] A. Voisard and H. Schweppe. Abstraction and decomposition in interoperable GIS. *Int. J. Geographical Information Systems*, 12, 1998.

[Wal91] D. R. F. Walker. Introduction to metadata in the geosciences. In D. Medykyj-Scott, I. Newman, C. Ruggles, and D. Walker, editors, *Metadata in the Geosciences*. Group D Publications, Loughborough, UK, 1991.

[WAS97] WASY GmbH, 1997. URL http://www.wasy.de.

[Whi81] M. White. N-trees: Large ordered indexes for multi-dimensional space. Technical report, United States Bureau of the Census, Statistical Research Division, Application Mathematics Research Staff, 1981.

[WHM90] M. F. Worboys, H. M. Hearnshaw, and D. J. Maguire. Object-oriented data and query modelling for geographical information systems. In *Proc. 4th Int. Symp. on Spatial Data Handling*, Zürich, 1990.

[WHSS92] L. Wicke, H.-D. Haasis, F. Schafhausen, and W. Schulz. *Betriebliche Umweltökonomie – Eine praxisorientierte Einführung.* Vahlen, Munich, 1992.

[Wid91] P. Widmayer. Datenstrukturen für Geodatenbanken. In G. Vossen and K.-U. Witt, editors, *Entwicklungstendenzen bei Datenbank-Systemen*, chapter 9, pages 317–361. Oldenbourg-Verlag, Munich, 1991.

[Wil96] R. Wilensky. Toward work-centered digital information services. *IEEE Computer*, 29(5), 1996.

[WK85] K.-Y. Whang and R. Krishnamurthy. Multilevel grid files. Technical report, IBM Research Lab., Yorktown Heights, N.Y., 1985.

[Wor92] M. F. Worboys. A generic model for planar geographical objects. *Int. J. Geographical Information Systems*, 6:353–372, 1992.

[Zad65] L. A. Zadeh. Fuzzy sets. *Information and Control*, 8:338–353, 1965.

[Zam88] P. Zamparoni. Feature extraction by rank-order filtering for image segmentation. *Int. J. Pattern Recognition Artif. Intelligence*, 2(2):301–319, 1988.

# Index

abstract class, 112
abstract data type, 53, 61–63, 112, 117, 118
abstraction, 112
access method, *see* multidimensional access method
adaptive k-d-tree, 72
adjacency query, 57
ADT, *see* abstract data type
ALBIS, 165
alert, 15
Alexandria, 208
AlfaWeb, 150
Altavista, 150
AML, 141
ARC/INFO, 49, 61, 129, 141, 144, 174, 175, 186
– and OpenGIS, 130
ArcView GIS, 141, 143, 144, 167, 175
– and OpenGIS, 130
– Internet Map Server, 151
– Network Analyst, 142
ASK, 150
Avenue, 141
AVHRR, 29, 30

BANG file, 78, 84–86, 90, 107
Bayes' rule, 14
Bayesian probability theory, *see* probability theory
BD-tree, 78, 86
belief, 19
bibliographic database, 146
binary large object, *see* long field
binary operator, 55–57
BLOB, *see* long field
bluesheet, 148
British Standard Institute
– specification for environmental management systems (BS 7750), 153
Brown is Green, 150

BSP tree, 72–74, 78, 103, 104
buddy tree, 78, 82–84, 90, 107
buffering, 142, 152
built-in object, 117
BV-tree, 110

CAEM, *see* environmental management information systems
cartographic object, 128
Catalogue of Data Sources, 150, 178, 187
CDS, *see* Catalogue of Data Sources
CEDAR, 150, 175
cell tree, 90, 102–105, 108, 109
certainty factor, 18
CHEMS, 158
CHEMTOX, 147
chromatography, 21–28
CIMI, 176
circumstantial information, 17, 36
class, 15, 112–114, 188, 197–199
– hierarchy, *see* inheritance
classification, 13–37
clipping, 90, 100–105, 107
CNR, *see* Italian National Research Council
CODATA, 175
Common Object Request Broker Architecture, 160, 168, 169
complex object, 115
composite object, *see* complex object
confusion matrix, 36, 37
containment query, 57
control area, 36
CORBA, *see* Common Object Request Broker Architecture
CORINE CDS, 187
Current Contents, 145
Cygnus Group, 150

Druck:          Strauss Offsetdruck, Mörlenbach
Verarbeitung:   Schäffer, Grünstadt